SHAPER HANDBOOK

Roger W. Cliffe &
Michael J. Holtz

Sterling Publishing Co., Inc. New York

Library of Congress Cataloging-in-Publication Data

Cliffe, Roger W.
Shaper handbook / by Roger Cliffe and Michael Holtz.
p. cm.
Includes index.
1. Shapers. 2. Woodwork. I. Holtz, Michael. II. Title.
TT186.C553 1990
684′.083—dc20 90-40300
CIP

3 5 7 9 10 8 6 4

Copyright © 1990 by Roger W. Cliffe and Michael J. Holtz
Published by Sterling Publishing Company, Inc.
387 Park Avenue South, New York, NY 10016
Distributed in Canada by Sterling Publishing
℅ Canadian Manda Group, P.O. Box 920, Station U
Toronto, Ontario, Canada M8Z 5P9
Distributed in Great Britain and Europe by Cassell PLC
Villiers House, 41/47 Strand, London WC2N 5JE, England
Distributed in Australia by Capricorn Ltd.
P.O. Box 665, Lane Cove, NSW 2066
Manufactured in the United States of America
All rights reserved
Sterling ISBN 0-8069-6798-6

ACKNOWLEDGMENTS

The *Shaper Handbook* represents the work, cooperation, and experience of many fine people. Without these people, the authors could not have completed the work.

Custom darkroom work was done by Bill Peters. Special thanks for making our photographs look better. Manuscript typing was done by Maggie Perry. She turned our longhand scriblings into a readable, coherent manuscript.

Commercial products, tools, and photographs were generously furnished by the following people and organizations:

Horst Biernath
Phillip Granquist

Bimex, Inc.
Cascade Precision Tool

Fred Garms & Steve Ekard
Barry Dunsmore
Steve "Buddy" Holley
Gene Sliga
Sandy Brady
Bill Possinger
Karen Cody & Karl Frey
Dick Riggins

Rick Weaver & David Vander Werff

DML, Inc. Deluxe Division
Freud, Inc.
Delta International

Machinery Corp.
Powermatic
Reliable Grinding
SCMI

Wisconsin Knife Works
L. A. Weaver Co., Inc.

METRIC EQUIVALENCY CHART

MM—MILLIMETRES CM—CENTIMETRES

INCHES TO MILLIMETRES AND CENTIMETRES

INCHES	MM	CM	INCHES	CM	INCHES	CM
⅛	3	0.3	9	22.9	30	76.2
¼	6	0.6	10	25.4	31	78.7
⅜	10	1.0	11	27.9	32	81.3
½	13	1.3	12	30.5	33	83.8
⅝	16	1.6	13	33.0	34	86.4
¾	19	1.9	14	35.6	35	88.9
⅞	22	2.2	15	38.1	36	91.4
1	25	2.5	16	40.6	37	94.0
1¼	32	3.2	17	43.2	38	96.5
1½	38	3.8	18	45.7	39	99.1
1¾	44	4.4	19	48.3	40	101.6
2	51	5.1	20	50.8	41	104.1
2½	64	6.4	21	53.3	42	106.7
3	76	7.6	22	55.9	43	109.2
3½	89	8.9	23	58.4	44	111.8
4	102	10.2	24	61.0	45	114.3
4½	114	11.4	25	63.5	46	116.8
5	127	12.7	26	66.0	47	119.4
6	152	15.2	27	68.6	48	121.9
7	178	17.8	28	71.1	49	124.5
8	203	20.3	29	73.7	50	127.0

YARDS TO METRES

YARDS	METRES	YARDS	METRES	YARDS	METRES	YARDS	METRES	YARDS	METRES
⅛	0.11	2⅛	1.94	4⅛	3.77	6⅛	5.60	8⅛	7.43
¼	0.23	2¼	2.06	4¼	3.89	6¼	5.72	8¼	7.54
⅜	0.34	2⅜	2.17	4⅜	4.00	6⅜	5.83	8⅜	7.66
½	0.46	2½	2.29	4½	4.11	6½	5.94	8½	7.77
⅝	0.57	2⅝	2.40	4⅝	4.23	6⅝	6.06	8⅝	7.89
¾	0.69	2¾	2.51	4¾	4.34	6¾	6.17	8¾	8.00
⅞	0.80	2⅞	2.63	4⅞	4.46	6⅞	6.29	8⅞	8.12
1	0.91	3	2.74	5	4.57	7	6.40	9	8.23
1⅛	1.03	3⅛	2.86	5⅛	4.69	7⅛	6.52	9⅛	8.34
1¼	1.14	3¼	2.97	5¼	4.80	7¼	6.63	9¼	8.46
1⅜	1.26	3⅜	3.09	5⅜	4.91	7⅜	6.74	9⅜	8.57
1½	1.37	3½	3.20	5½	5.03	7½	6.86	9½	8.69
1⅝	1.49	3⅝	3.31	5⅝	5.14	7⅝	6.97	9⅝	8.80
1¾	1.60	3¾	3.43	5¾	5.26	7¾	7.09	9¾	8.92
1⅞	1.71	3⅞	3.54	5⅞	5.37	7⅞	7.20	9⅞	9.03
2	1.83	4	3.66	6	5.49	8	7.32	10	9.14

CONTENTS

1
INTRODUCTION TO THE SHAPER

How a Shaper Works

Many people believe that any device which will cut shapes or profiles in wood is a shaper (Illus. 1-1). This is not the case, however. Many routers and other tools with high-speed motors can cut shapes and profiles. They cannot, however, cut several profiles or shapes in a single pass and cannot cut from two directions. This is what a shaper does, and does well!

The shaper is a stationary woodworking machine designed to cut decorative mouldings and joinery. The piece on the shaper that does the actual cutting is called, aptly enough, the cutter. Cutters are mounted on the shaper spindle or arbor.

The work is supported on the shaper table while it is being shaped. In most cases, stock is controlled by the fence and table while it is being shaped (Illus. 1-2). In some cases, stock is controlled with a mitre gauge (Illus. 1-3) or a jig or fixture (Illus. 1-4 and 1-5). This approach is

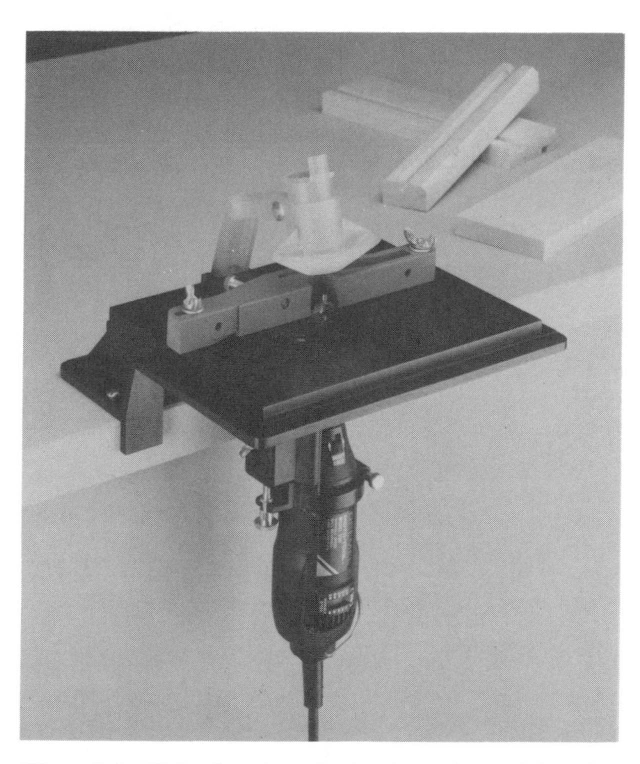

Illus. 1-1. This shaping device is not considered a shaper. However, it will cut many profiles on light stock. (Photo courtesy of Dremel)

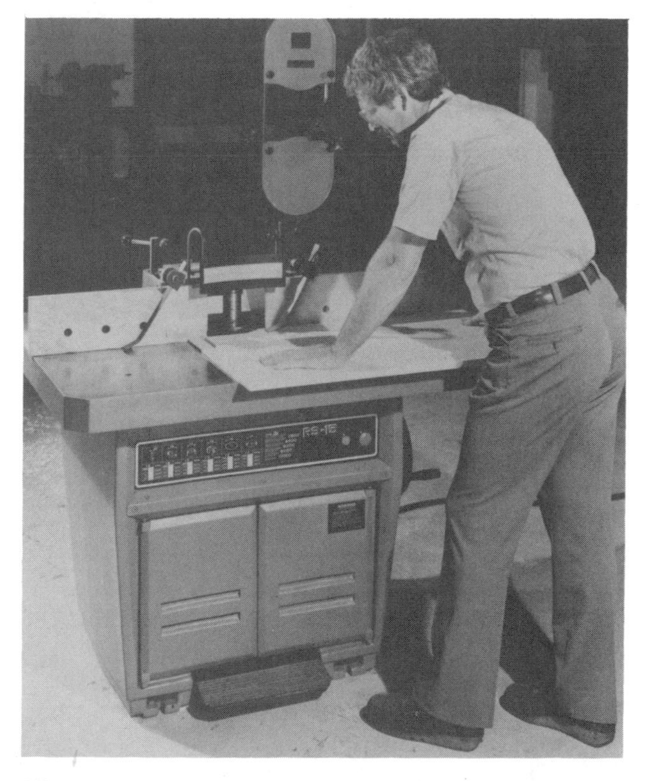

Illus. 1-2. Stock is usually controlled by the fence and table. Setups vary according to the cutter and intended purpose.

Illus. 1-3. The stock can also be controlled with the mitre gauge. The mitre gauge is usually used to control stock whose end grain is being shaped.

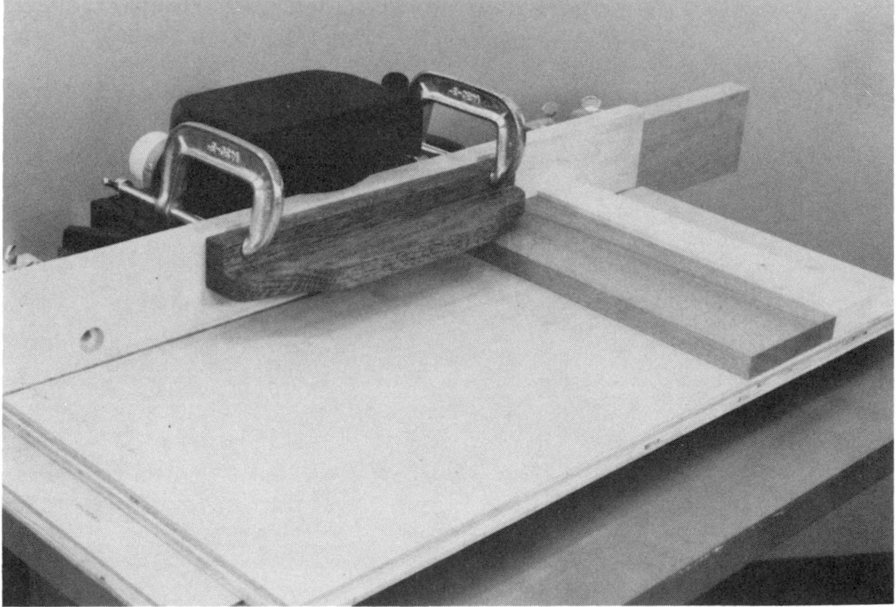

Illus. 1-4. This jig is being used to control end grain. It slides on a plywood auxiliary table. The depth of cut is controlled by the fence.

used for shaping irregularly shaped parts or parts which require additional control. The depth of the cut is controlled by the jig or fixture. The jig or fixture either rubs against a depth collar that is mounted on the shaper spindle (Illus. 1-6), or it rides along the fence (Illus. 1-7 and 1-8).

The spindle assembly (Illus. 1-9) is the most important part of the shaper. It is driven by an electric motor. A belt connects the motor to the spindle (Illus. 1-10). The spindle then turns the arbor and shaper cutter (Illus. 1-11 and 1-12), and the shaper cutter cuts the desired joint or profile.

The diameter of the spindle or arbor ranges from ½ to 1¼ inches. The larger the spindle diameter, the less the spindle deflects and vibrates. This is because a spindle with a larger diameter is more rigid. It resists deflection when stock is fed into the shaper cutter.

Some shapers have interchangeable spindles or arbors. This allows more versatility. The user can mount cutters with different spindle diameters on the shaper. There are even some shapers that offer a router collet-type spindle as an accessory for the shaper. This allows the shaper to use router bits. Router bits are generally less expensive than shaper cutters. This makes the shaper more versatile and less expensive to use.

Types of Cut Made by the Shaper

There are many different cutters available for shapers. They are sold individually or in sets. Individual cutters can be stacked on the spindle

Illus. 1-5. Vertical shaping can be controlled with this jig. Chapter 6 explores some applications for this jig.

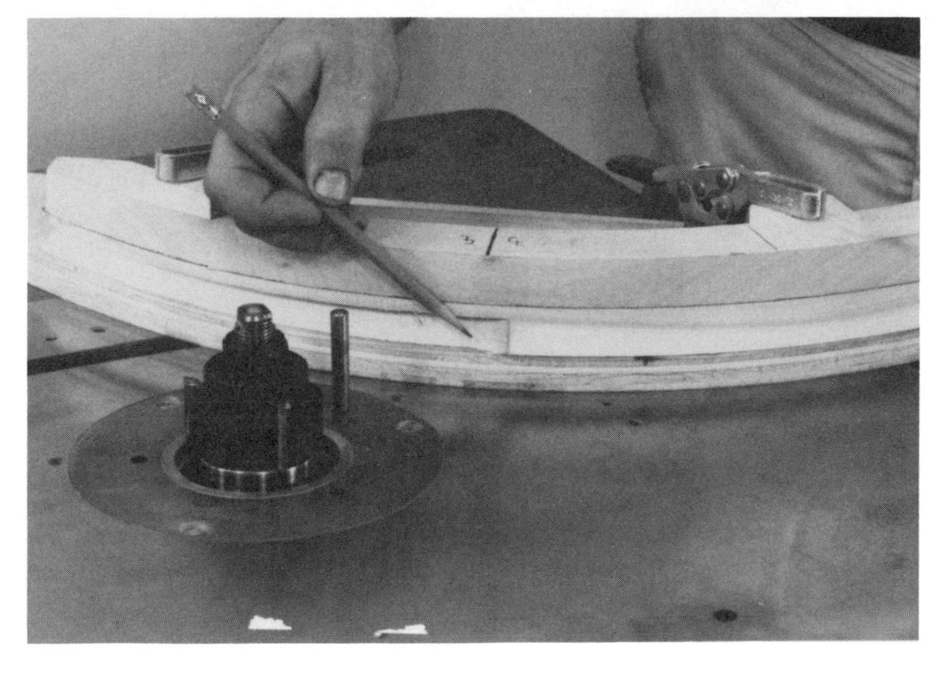

Illus. 1-6. The depth collar under the shaper cutter rides along the base of this jig. The depth collar and jig control the path of the cut.

Illus. 1-7. This series of jigs and fixtures is used on shapers to make all the parts for a cabinet door. (Photo courtesy of L.A. Weaver)

Illus. 1-8. This Panelcrafter™ jig, a commercial accessory, can be used with the fence or a depth collar. It has pneumatic clamps to hold stock in position while it is being shaped.

to make decorative cuts. Cutter sets are sold for specialty jobs such as mating cope-and-stick joinery on cupboard or entry doors. Since the cutters are offered as a set, the fit between the parts can be assured.

Shaper cutters can be either solid or assembled. Solid cutters are of one piece (Illus. 1-13–1-17). An assembled cutterhead consists of shaper cutters that are fastened onto the shaper head (Illus. 1-18 and 1-19). You can use an assembled cutterhead to cut numerous shapes simply by changing the cutters.

Cutters cannot be stacked on an assembled cutterhead, but you can discard them when they become dull. Many assembled cutters are offered as door kits or other specialty cutter kits (Illus. 1-20).

Whatever types of cutter you use, make sure that they have been designed for the spindle speed of your shaper. *Operating a cutter above its intended speed could cause serious injury or damage.*

Illus. 1-9. This spindle assembly is mounted in the shaper. It has two sets of ball bearings to reduce wear and make shaping smoother.

Illus. 1-10. The belt connects the drive pulley to the driven pulley on the shaper assembly.

Shaper Use in the Shop

The shaper is a workhorse in either the workshop or production shop. It can be used to make short to medium runs of moulding or millwork. Cabinetmakers frequently use the shaper for producing cabinet doors and drawers. Door and drawer setups are quick on the shaper, and the pieces are uniform in shape. In fact, some shaper manufacturers offer entire door or drawer systems (Illus. 1-21). These are a group of shapers used to make all the cuts required to produce a door or drawer. These machines are always ready to produce doors or drawers; this reduces set-up time and bottlenecks on other machines such as the table saw or radial arm saw.

Mill shops also use shapers to fabricate exterior and interior house doors (Illus. 1-22—1-24). These are the doors which contain raised panels or panes of glass. In fact, shapers are the most productive machine for house door construction. This is because when closely toleranced cutters are mounted on an accurately machined shaper, the results are a mating cope-and-stick (male and female) joint that fits together perfectly (Illus. 1-25).

The work produced by the shaper can be as simple as porch balusters, as beautiful as entry doors on a building, and as ornate as the communion rail or altar in a church. As you work with the shaper, you will come to realize how many jobs can be completed with it. The versatility of the shaper is only limited by your imagination and supply of cutters.

Illus. 1-11. The arbor holds the shaper cutter to the spindle assembly.

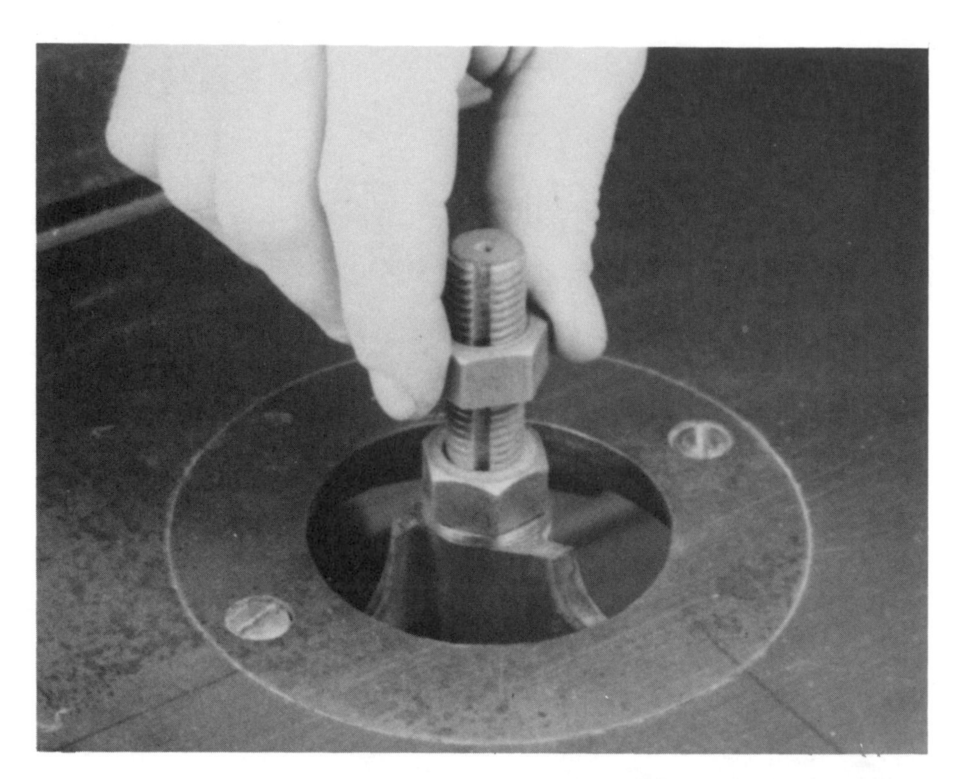

Illus. 1-12. Shaper cutters are always held to the arbor with threaded fasteners.

Illus. 1-13. These one-piece cutters have three cutting edges. They round the edge of a piece of stock or shape a bead in the face or edge of the stock.

Types of Shaper

When shapers were first invented, there was only one type. It was a vertical spindle machine which used a rotary cutting concept. This concept can be traced to Samuel Bentham. The shaper was patented by Mr. Bentham in England in the 1790s. It was used in the shipbuilding industry.

The shaper of the 1790s was made chiefly of wood, with a metal spindle and cutterhead. Modern shapers are made entirely of metal. They have ball-bearing spindles and safer, more accurate cutters. Industrial shapers usually have

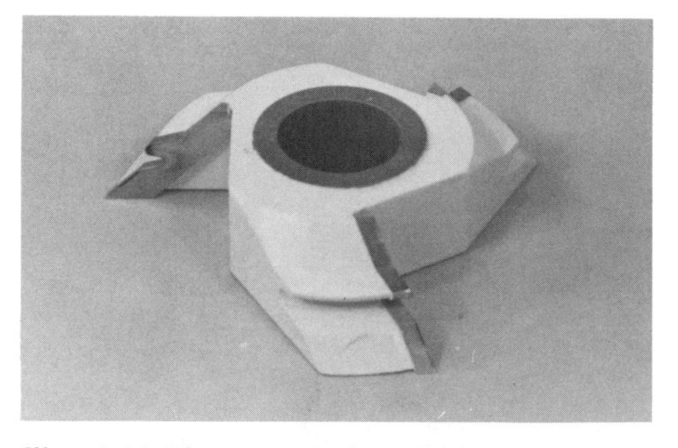

Illus. 1-14. This cutter makes a lock mitre joint. It shapes the entire edge of a piece at a 45-degree angle. This cutter requires great power to make the cut.

Illus. 1-15. The large diameter of these raised panel cutters requires that spindle speed be reduced for proper shaping.

Illus. 1-16. The raised panel cutter turns faster at its outer edge. If the spindle speed is too high, the cutter is likely to burn near the center of the workpiece.

Illus. 1-17. This tall cutter shapes wide or thick parts. It requires great power to use the entire cutter in a single cut.

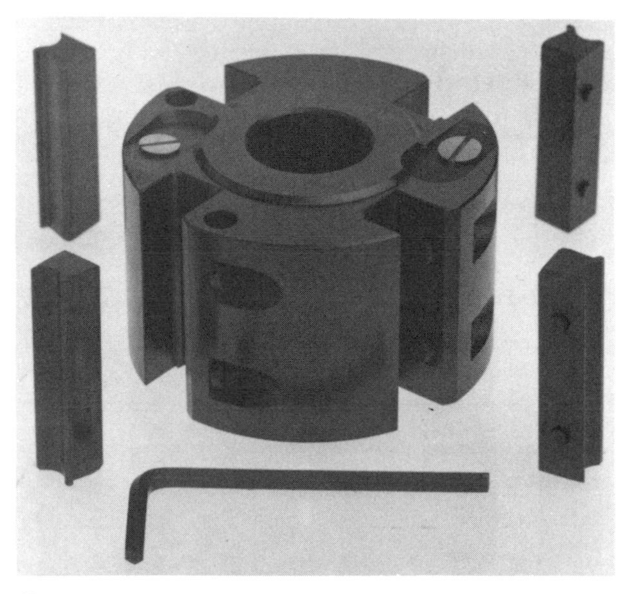

Illus. 1-18. This assembled cutterhead holds four cutters. The Allen wrench is used to secure them. (Photo courtesy of BIMEX)

massive cast-iron bases and tables which increase the rigidity of the machine.

There are several types of shaper available on the marketplace today. One of the chief distinctions between these shapers is the type of spindle they have. Therefore, they will be discussed under the following heads only according to the classification of their spindles.

SINGLE-SPINDLE SHAPERS

The most common type of shaper has a single spindle. A shaper with a single spindle usually has a reversible motor so that the spindle may turn clockwise or counterclockwise. The spindle may be located in the center of the table, or it may be closer to the rear of the table. Manufacturers who locate the single spindle towards the rear of the table contend that this provides more working table for irregularly shaped parts.

Most single-spindle shapers offer interchangeable spindles. Interchangeable spindles allow greater versatility. Cutters with any size spindle bore can be used on an interchangeable spindle shaper.

TILTING SPINDLE SHAPERS

Some single-spindle shapers have a tilting spindle. This increases the shaper's versatility. You can change the shape made by any cutter simply by tilting the spindle. The number of cutting profiles attainable with a single cutterhead is increased dramatically.

When the spindle is tilted, more of the cutterhead can be engaged with the work. In other words, you can increase the depth of cut when you tilt the cutter.

Always check the depth of cut when tilting the spindle. The cut should not be too deep. Also, you may have to change the position of the fence when you tilt the spindle. As you read the chapters on shaper safety and setup, you will better understand these setup precautions.

Illus. 1-19. This assembled cutterhead uses disposable cutters. There are multiple cutting edges on these cutters.

DOUBLE-SPINDLE SHAPERS

Double-spindle shapers usually have a table twice as large as a single-spindle shaper. Two

Illus. 1-20. Two cutters are required to make a tongue-and-groove joint. If the cutters are from the same manufacturer, they should fit snugly against each other.

Illus. 1-21. This set of shapers is used to make doors without changing any setup. If doors are your specialty, this may be the best approach. (Photo courtesy of Powermatic)

Illus. 1-22. This shaper has been set up to shape raised panels. It uses a vertical spindle cutter that reduces the amount of horsepower needed.

spindles project through the table as if two single spindle machines have been placed side by side.

Double-spindle machines can be used in three different ways. First, a fence can be used between the spindles. This means that a different cut can be made by each spindle. A piece of stock can be shaped twice as it crosses the table of a double-spindle shaper.

Second, a fence can be used and identical cutters can be mounted on both spindles. The depth of cut will be deeper on the second spindle. As the work crosses the table, two consecutive cuts will be made on the work.

Third, two cutters that have the same profile but are designed to turn in opposite directions can be mounted on the spindles. No fence is used with this setup. The depth of cut is controlled with depth collars or a template. The spindles rotate in opposite directions, and parts with an irregular configuration are shaped. This setup allows shaping to be done with the grain regardless of the grain orientation of the part.

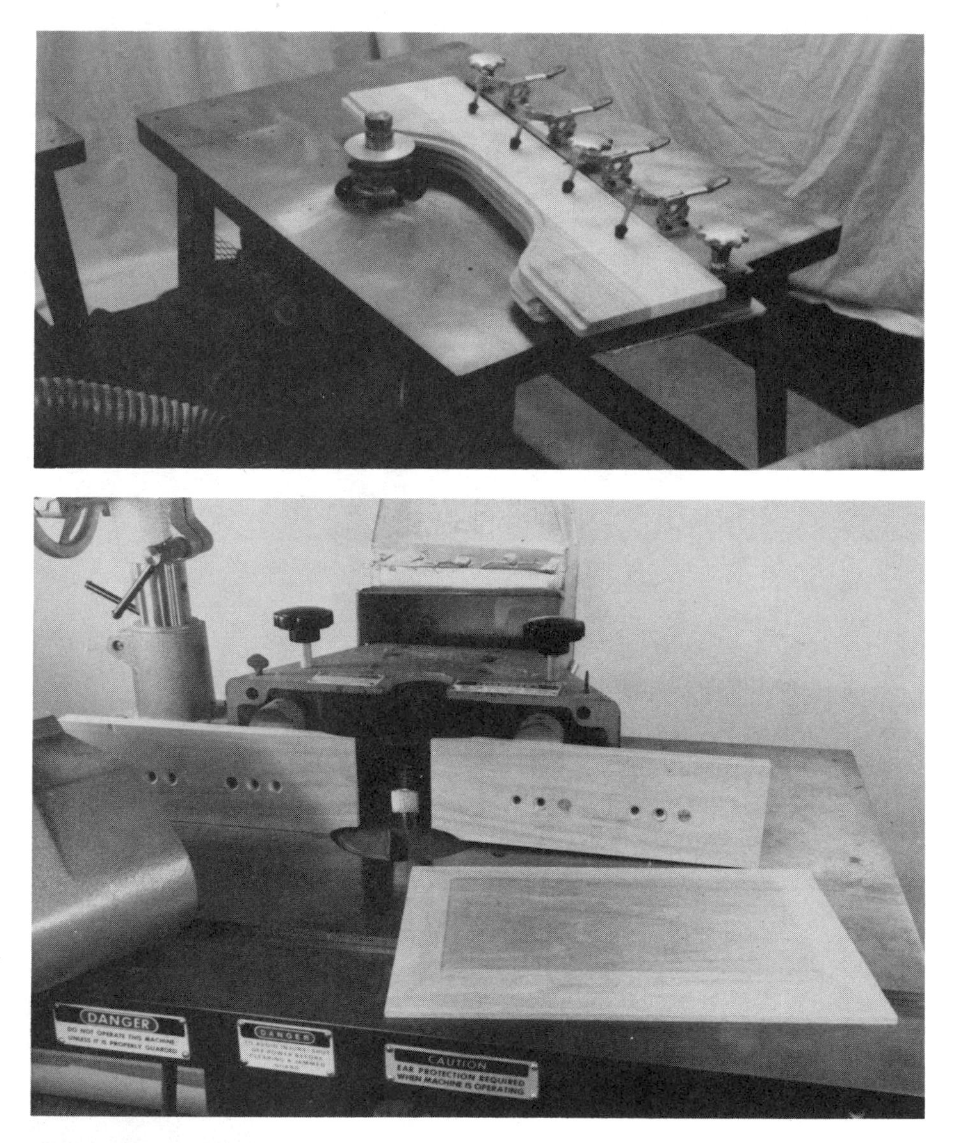

Illus. 1-23. This template is used to control the shaping of a house door part.

Illus. 1-24. This power feeder is set up to shape raised panels safely and efficiently.

Illus. 1-25. The cope-and-stick joint is a two-part joint. The rails (the horizontal parts of the door) are coped. The stick cut is made on the inside edge of the rails and stiles (vertical parts of the door).

Double spindle shapers are seldom found in the small cabinetshop or millshop. They are found in industrial woodworking facilities where pieces with irregular configurations are commonly shaped. They can be used to their best advantage in this type of setting.

HORIZONTAL SPINDLE SHAPERS

Most woodworkers think of the shaper as a stationary power tool with a horizontal table and a verticle spindle. During the 1980s, a shaper with a horizontal spindle was developed and patented. This shaper closely resembles a jointer. It was designed chiefly to shape all the parts needed to make a raised panel door (Illus. 1-26).

Fences control the workpiece the same way the table on a vertical spindle shaper does (Illus. 1-27–1-29). The fence on the horizontal spindle shaper also acts as a guard. Cutters on a horizontal spindle shaper are also installed much differently than cutters on a vertical spindle shaper. Consult the owner's manual for specifics.

Shaper Size

There are a number of factors which are used to determine the size and capabilities of a shaper. Understanding how these factors contribute to its performance will help you use the machine more effectively.

SPINDLE SIZE

The most important element in determining the size of a shaper is the diameter of its spindle. Spindles range in diameter from ½ to 1¼ inches (12–31 mm). The larger the spindle's diameter, the less tendency it has to deflect during a heavy cut. This means less vibration, less wear on the cutter, and a smoother cut on the work.

Small shapers are designed for light cuts, so a ½-inch spindle is adequate. Shapers designed for heavy cuts fed by a power feeder are subject to increased deflection. A spindle diameter of 1 inch or greater would be best in these cases.

CAPACITY

Another criterion that helps determine the size of a shaper is its capacity—sometimes known as "capacity under the nut." This refers to the overall height of the cutters and collars or shims that may be stacked on the spindle for a single cut. It also presumes that all the threads on the spindle nut will be engaged with the spindle.

The capacity of a shaper is important because it determines the maximum width or thickness that may be shaped in a single pass. For millwork shops, the capacity of a shaper is as important as the diameter of the spindle.

Illus. 1-26. This horizontal spindle shaper is designed to shape raised panels. (Photo courtesy of Lancaster Machinery).

Spindle Travel

Spindle travel is the total distance which the spindle can be elevated or lowered. This distance varies from one shaper to the next, but is directly proportional to the capacity under the nut. Spindle travel affects how the shaper can be adjusted and set up. Decreased spindle travel means the shaper setup must be made more carefully. You cannot rely on spindle travel to make course adjustments.

TABLE SIZE

The size of the table on a shaper is usually directly proportional to the diameter of its spindle and to its horsepower (Illus. 1-27–1-29). Some shapers offer a table extension that can be bolted to the front or back of the table. A larger table is desirable if you are going to edge- or face-shape long straight or curved parts.

In some cases, however, a large front table can be a hindrance. It can make it difficult for the

Illus. 1-27. This ½-inch spindle shaper has a relatively small table.

Illus. 1-28. This ¾-inch spindle shaper has a larger table and cabinet base.

operator to get near the workpiece and mitre gauge (Illus. 1-30 and 1-31). As a result, some shapers have a rolling table connected to the front or side of the shaper (Illus. 1-32). This table provides support when you are shaping end grain. For other operations, you can turn the fence to face the opposite side of the shaper.

A large rear table may be desirable for mounting a power feeder base. In order for the power feeder to perform correctly, it should be mounted securely to the shaper. The base must be mounted in an out-of-the-way spot. This is difficult on a shaper with a small table. It would be difficult to mount a power feed base on a bench-type shaper.

MOTOR

The motor is another important determinant of shaper size. Generally, shaper motors turn at 3,450 rpms (revolutions per minute). The motor is connected to a drive pulley. The drive pulley is connected to the spindle pulley with a belt. The drive pulley is usually larger in diameter than the spindle pulley. This increases the speed of the spindle.

On some shapers, the spindle has two or more pulleys of different diameters. This means that you can vary the speed of the spindle according to the type of cut, diameter of the cutter, the

Illus. 1-29. This 1¼-inch spindle shaper has a massive cast-iron table and combination rolling table and table extension.

Illus. 1-30. The front-table extension controls the operator's position during this shaping operation.

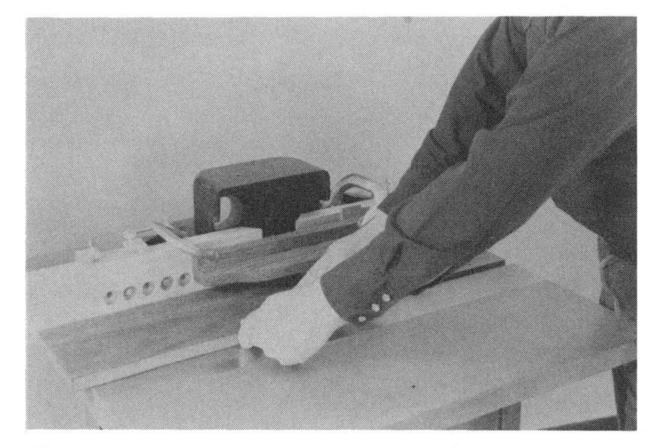

Illus. 1-31. The extension table makes it difficult to get close to the shaper. The extension table is useful when you are using jigs and fixtures.

number of cutting edges on the cutter, and the hardness of the stock. Spindles typically range in speed from 7,000–10,000 rpm.

Generally, the larger the diameter of the cutter, the slower the speed of the spindle should be. This is because the spindle's peripheral speed increases greatly as the diameter of the cutter increases. A spindle with a high peripheral speed can mean that you will end up burning the stock instead of cutting it.

Power feeders must also be adjusted so that they are compatible with the speed of the spindle and the diameter of the cutter (Illus. 1-33 and 1-34). If the feed speed is too slow, the stock may burn and productivity will be reduced. If the feed speed is much greater than the speed of

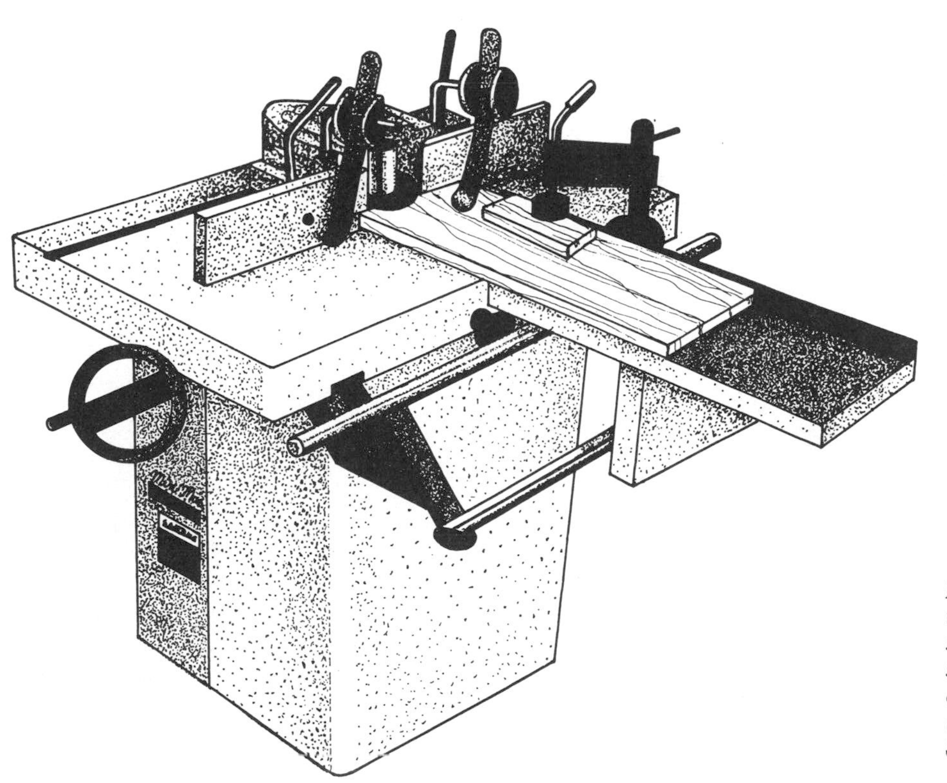

Illus. 1-32. This rolling table supports heavy stock. It also controls the stock with a clamp during end-grain shaping. (Drawing courtesy of SCMI/Mini-Max)

Illus. 1-33. Power feeders should be adjusted to a feed speed that is compatible with the speed and diameter of the cutter. This power feeder has two horsepower.

the spindle, the quality of the cut will be affected. There is likely to be tearout or large mill marks.

HORSEPOWER

The horsepower of the motor also plays a part in determining the capability of a shaper. An underpowered shaper usually means less efficiency and increased labor. Shapers range in horsepower from a minimum of one to ten horsepower. Horsepower is usually directly proportional to spindle diameter and table size.

For door shaping, the shaper should have a 1½ horsepower motor. A motor with less than three horsepower may not be capable of cutting copes or raised panels in a single cut (Illus. 1-35 and 1-36). When a shaper cannot form the profile in a single cut, labor is wasted on a second cut.

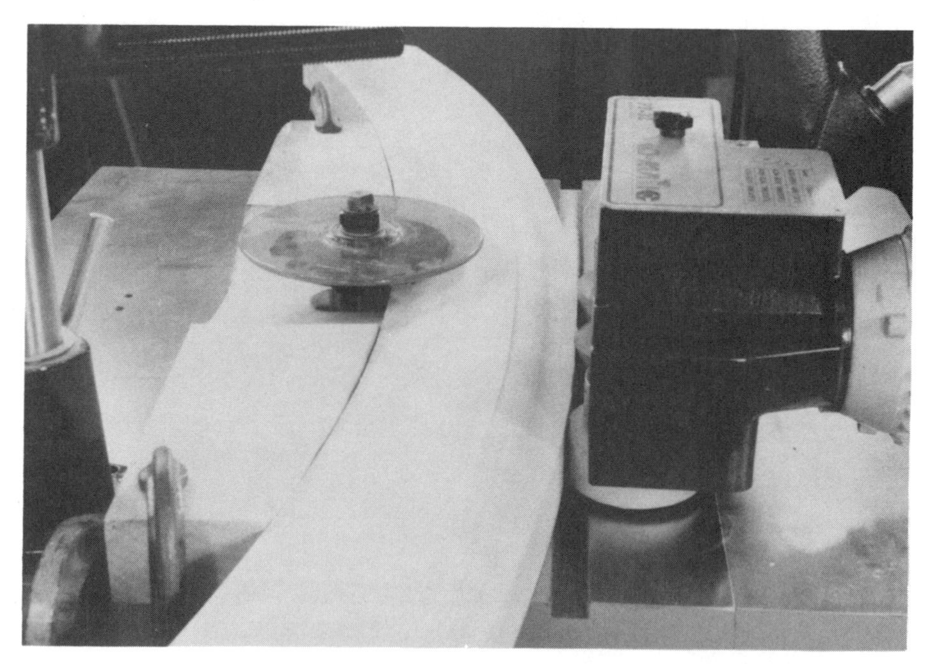

Illus. 1-34. This ¼-horsepower power feeder is being used in the horizontal position with a curved fence. It is being used with a 1½-horsepower shaper. The cutter is a raised panel cutter that turns at 7,000 rpm.

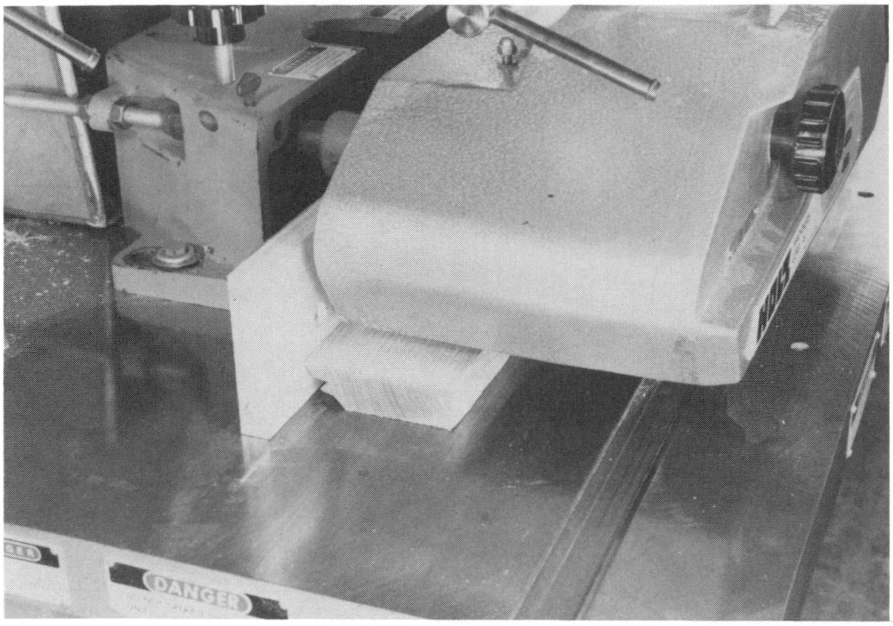

Illus. 1-35. This 3-horsepower feeder is used with a 5-horsepower shaper. The cutter turns at 8,000 rpm.

Power feeders must have adequate power to feed stock into the shaper. The power feeder and shaper must be compatible but not equal in horsepower. If a power feeder is used, 1/6–1/2 horsepower is required as a minimum. If you are cutting large profiles using a power feeder, then 3–5 horsepower is considered the minimum (Illus. 1-37).

Shapers can operate on voltages of 110 and up. The most common voltages are 110 and 220. A motor using 110 volts and 12 to 15 amperes would have about a two-horsepower rating. Larger motors will use 220 volts and 8–20 amps.

Some 220-volt machines run on a single-phase current, while others run on a three-phase current. Three-phase power is most commonly found in industrial parks and commercial zones. It is rarely found in residential areas.

Large horsepower shapers are only suited to three-phase power and are used only in industrial applications.

Illus. 1-36. The mating part of the lock mitre is cut with the power feeder in the vertical position.

Illus. 1-37. You have to use a 5-horsepower shaper with a 3-horsepower feed to raise both sides of this panel in one cut.

Illus. 1-38. This Mini-Max shaper is designed for the cabinet and mill shop. (Drawing courtesy of Mini-Max/SCMI)

Shaper Overview

The shapers shown in Illus. 1-39–1-49 are representative of those shapers being marketed today. They are presented here to give you an idea of the types of shapers and accessories that are available commercially. Remember to consider all the factors explored in this chapter when determining which shaper to buy.

Illus. 1-39. This Delta cabinet shaper is available in 1½, 3, and 5 horsepowers.

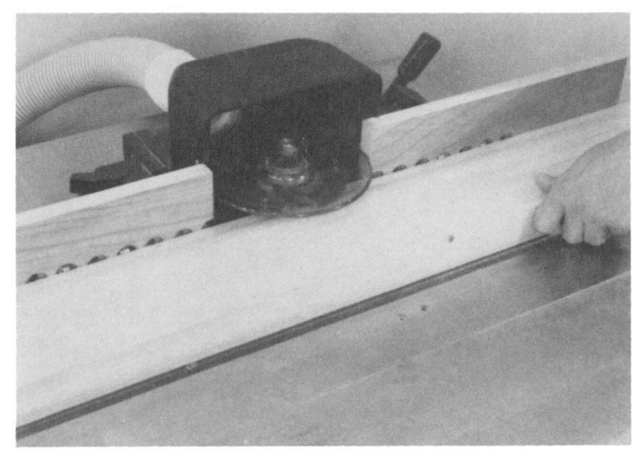

Illus. 1-40. The Delta shaper has an iron dust hood for chip collection.

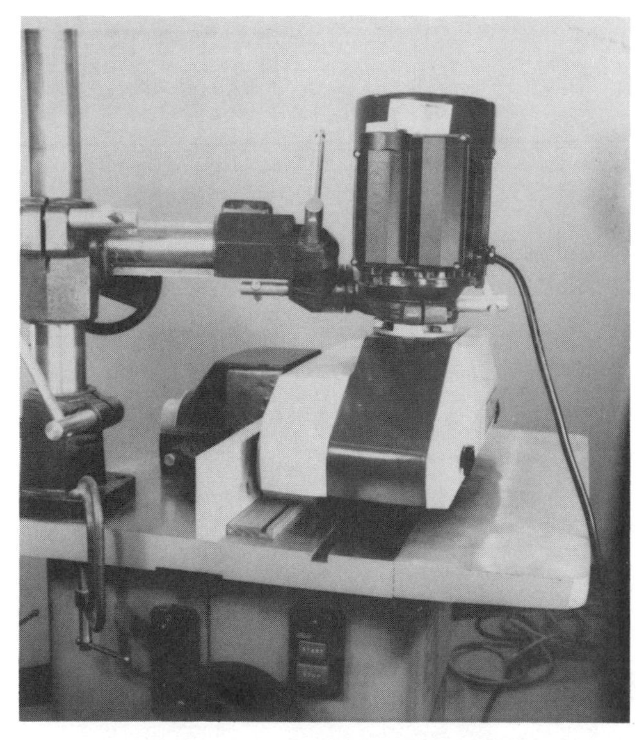

Illus. 1-41. A 2-horsepower power feeder has been mounted onto this 1½-horsepower shaper. A lighter power feeder could be used for most applications.

Illus. 1-42. This Powermatic cabinet shaper is available in 3 and 5 horsepowers. It comes with a complete guard system. It is sometimes used as one of a series of shapers in a door system (See Illus. 1-21.) (Photo courtesy of Powermatic)

Illus. 1-43. This Cascade cabinet shaper has a light base. It has two spindles that have ½- and ¾-inch diameters.

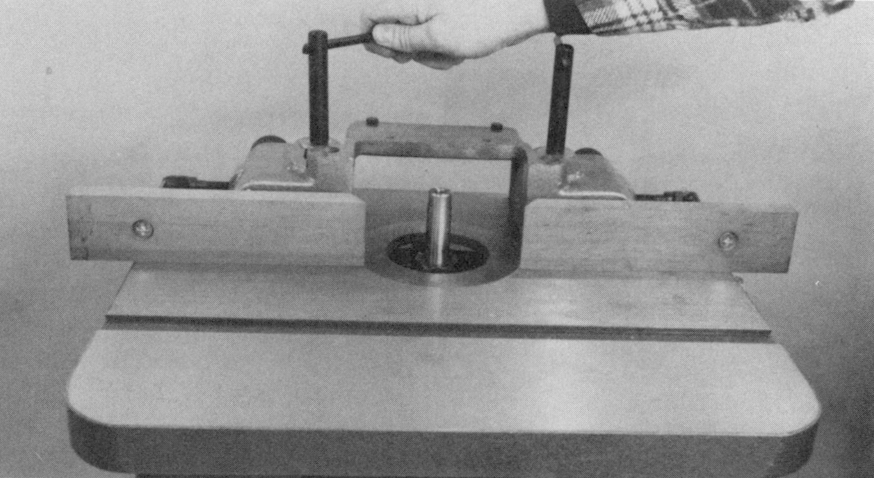

Illus. 1-44. The fences on this shaper are easy to adjust and lock.

Illus. 1-45. The smaller shaper will cut raised panels if a vertical cutter is used.

Illus. 1-46. This vertical raised-panel cutter requires less power to shape the work. This shaper cannot handle cutters greater than 3 inches in diameter. It will, however, accommodate router bits.

Illus. 1-47. This is the Powermatic Artisan Shaper. It has a collet for router bits and a ½-inch spindle. (Photo courtesy of Powermatic)

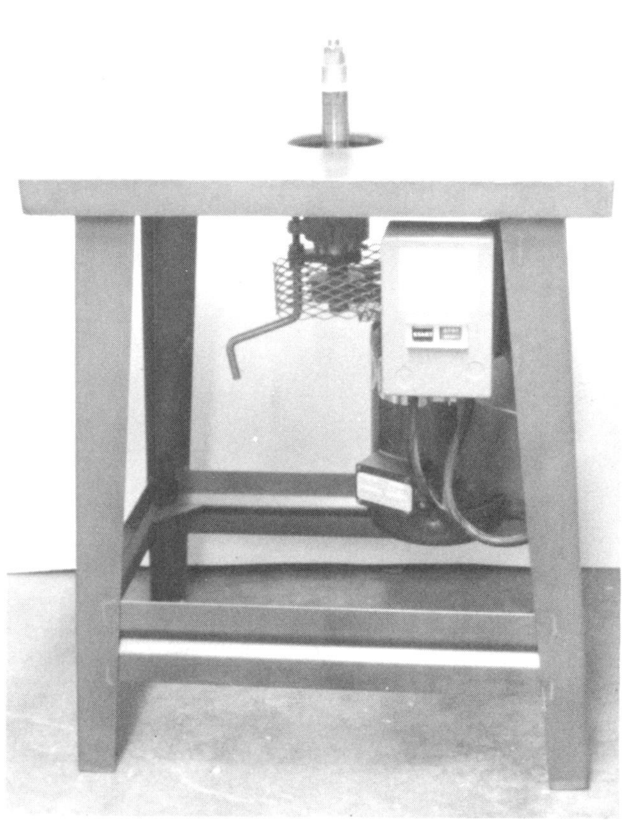

Illus. 1-48. This L. A. Weaver cabinet shaper has a large table, and a clean and simple design. (Photo courtesy of L. A. Weaver)

Illus. 1-49. Several Weaver shapers can be used together for door or drawer shaping. (Photo courtesy of L. A. Weaver)

2
CONTROLS AND ACCESSORIES

Shaper controls and accessories vary from manufacturer to manufacturer. This chapter examines the controls and accessories commonly used with shapers. The information you learn here will help you to better understand the terms and techniques explored in the following chapters.

Terminology and Nomenclature

As discussed in Chapter 1, shaping machines come in a number of different styles. The most common type of shaper is the single-spindle

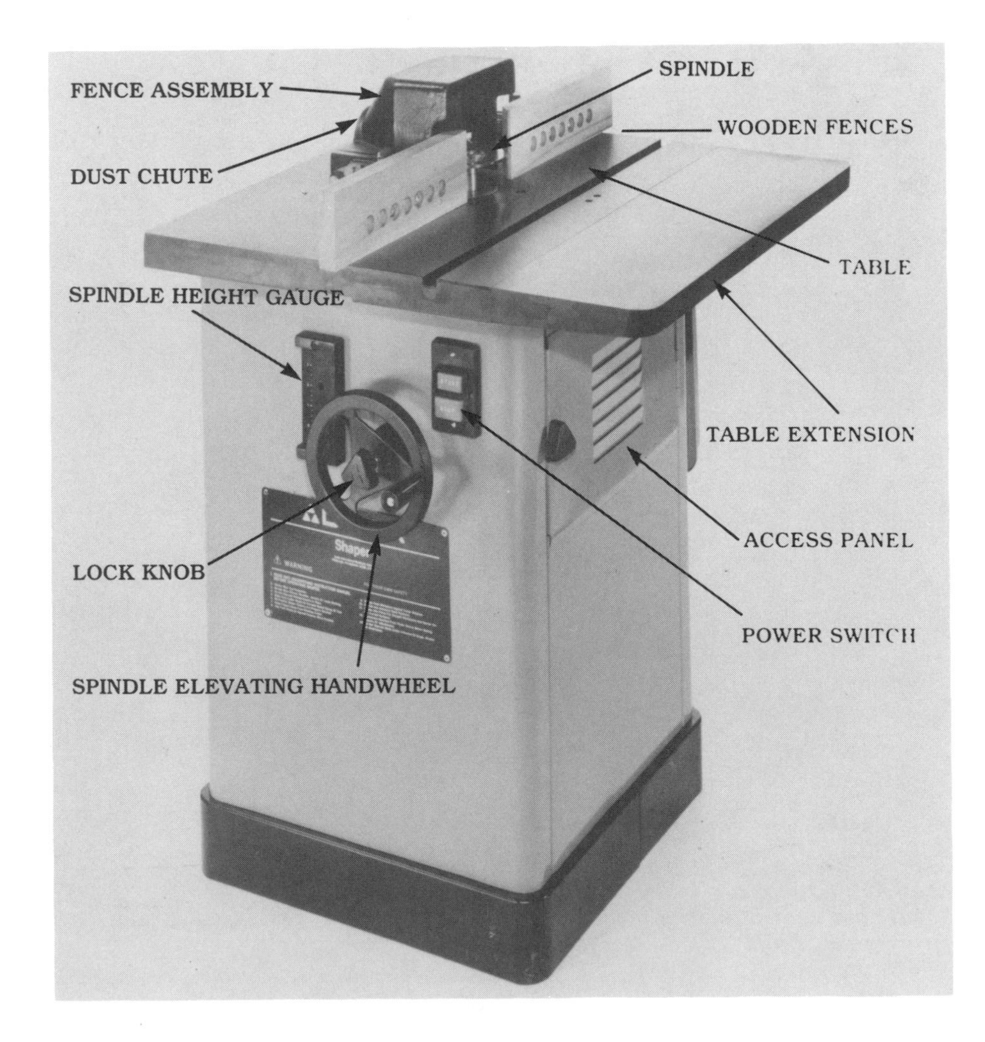

Illus. 2-1. This wood shaper is known as a cabinetmaker's shaper. It is used in job shops and cabinet shops that have intermediate runs of moulding or doors. Larger shapers are used for production runs. (Photo courtesy of Delta International Machinery Corporation)

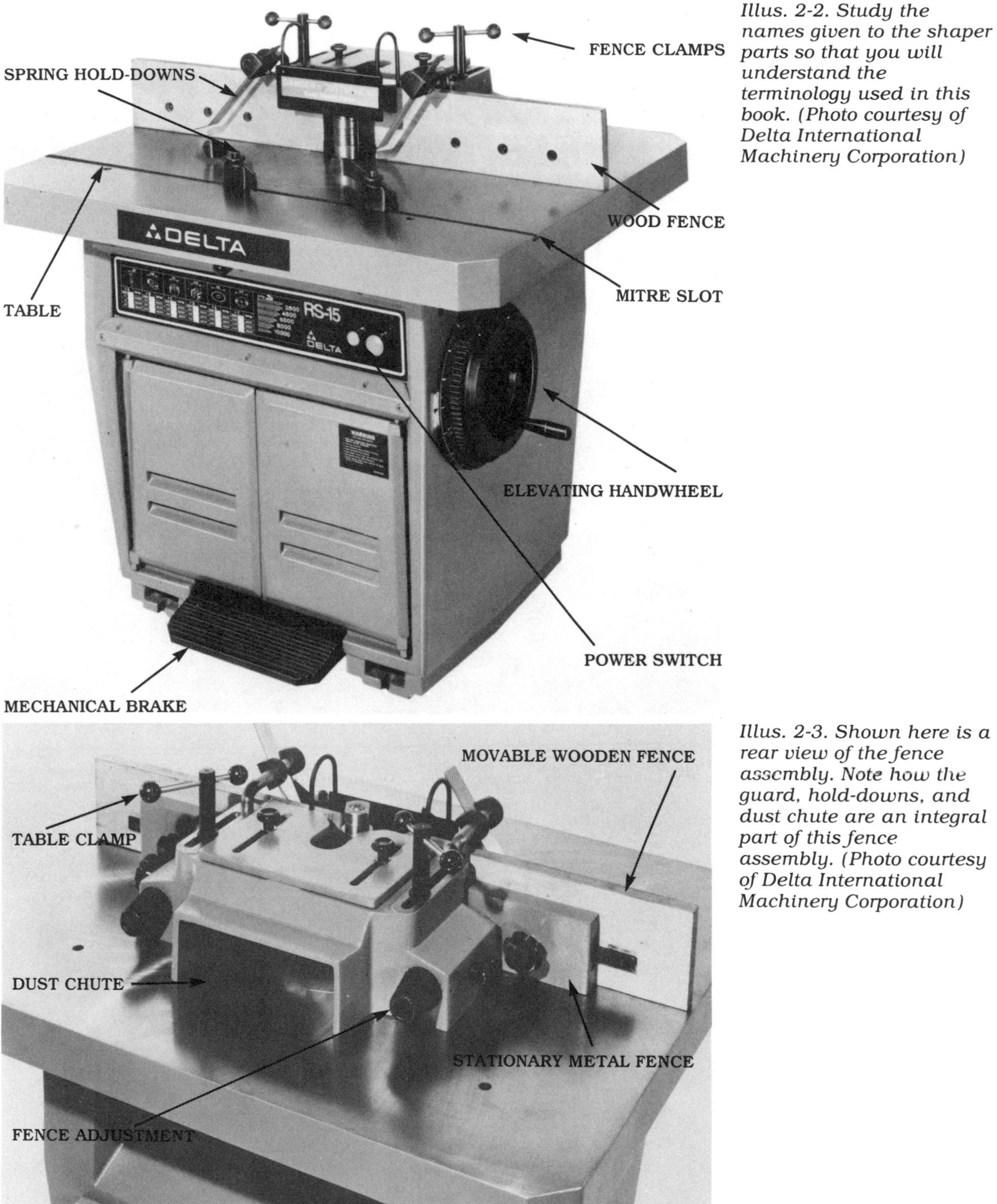

Illus. 2-2. Study the names given to the shaper parts so that you will understand the terminology used in this book. (Photo courtesy of Delta International Machinery Corporation)

Illus. 2-3. Shown here is a rear view of the fence assembly. Note how the guard, hold-downs, and dust chute are an integral part of this fence assembly. (Photo courtesy of Delta International Machinery Corporation)

shaper. The parts of this shaper are shown in Illus. 2-1 and 2-2 and discussed below. The terminology and nomenclature are common to other shapers discussed in this book. Study Illus. 2-1 and 2-2 to familiarize yourself with the terminology and parts.

Controls

POWER SWITCH

The most frequently used control on any shaper is the power switch. In most cases, the shaper comes with a reversible power switch. This allows the motor to be energized for either clockwise or counterclockwise rotation (Illus. 2-4).

Illus. 2-4. The switch on this shaper has a forward and reverse position. Make sure that you know which way the spindle is turning when it's in the forward or reverse position. Never reverse the direction of the spindle while the shaper is under power. Always allow the spindle to come to a complete stop before changing directions.

Some switches are push-button switches (Illus. 2-5). In some cases, the buttons are marked forward, reverse, and off. In other cases, the push buttons are marked on and off; a separate switch controls the direction of the cutter rotation (Illus. 2-6).

Other switches have a three-position handle. When the handle is turned to the clockwise position, the spindle should turn clockwise. When the handle is turned to the counterclockwise position, the spindle should turn counterclockwise. In some cases, the rotation of the power switch does not agree with spindle rotation. Always check this before setting up any shaper.

Many industrial shapers also have a brake. The brake can be used to shut off the shaper at any time (Illus. 2-7). When the shaper has a brake, it usually also has a spindle lock. When the brake is depressed, the spindle lock engages with the spindle. This holds the spindle securely while you remove or replace the cutters.

Many industrial shapers also have a "kill" button (Illus. 2-8). The "kill" button cuts the electrical power to the motor. This is a desirable feature in the event of an emergency or problem situation.

Some electrical systems have overload protection at the switch or on the motor. The overload switch protects the motor from overheating or burnout. Low-voltage switches are sometimes used on shapers. A low-voltage switch has only 24 volts. This prevents you from getting electrical shock when you touch the switch. The 24-volt switch also disconnects automatically if there is a power failure. This means that when power is restored, the switch is off.

Read your owner's manual to learn what features your shaper's electrical system offers. This will enable you to operate your shaper more efficiently.

Illus. 2-5. The power switch on this machine has a lockout feature. When the lockout button is in position, the shaper cannot be started. This feature is ideal for setup and maintenance work.

Illus. 2-6. This shaper has an independent control inside the machine which allows you to reverse the direction the cutter rotates at. This control is positioned in such a way that it is difficult to accidentally flick it.

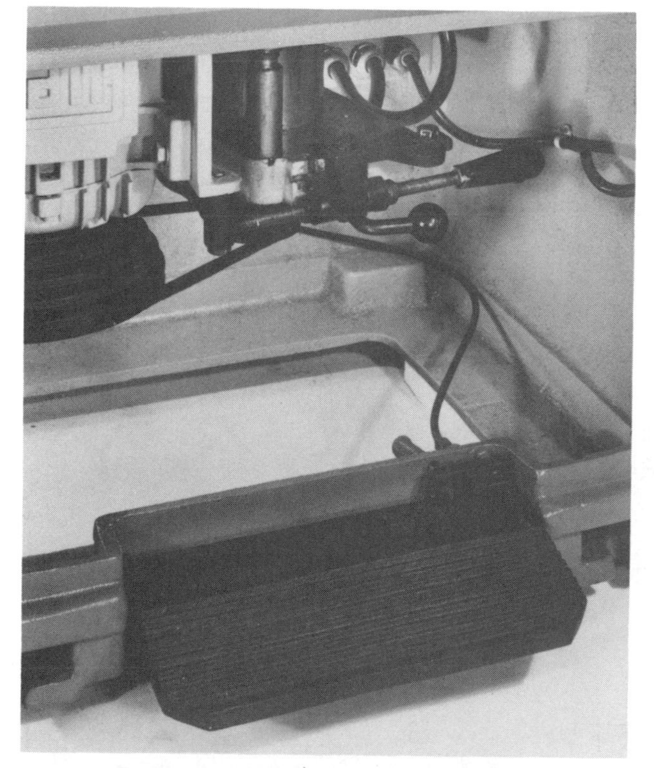

Illus. 2-7. The mechanical brake and spindle lock provide fast braking action and prevent the spindle from moving when you are setting the shaper up. (Photo courtesy of Delta International Machinery Corporation)

Illus. 2-8. The brake on this machine is an automatic shut-off brake. This means that if you push it, the machine shuts off and the spindle comes to a complete stop. There is also a kill button on the control panel. It, too, shuts off the machine instantly.

Illus. 2-9. The elevating handwheel raises and lowers the spindle.

Illus. 2-10. The elevating handwheel on the side of this shaper controls the height of the spindle.

SPINDLE-ELEVATING HANDWHEEL

The spindle elevating handwheel is used to position the cutter (Illus. 2-9 and 2-10). Cranking the handwheel clockwise usually elevates the spindle. Generally one revolution of the elevating handwheel will raise the spindle $1/32$ inch.

When the handwheel is turned, the yoke is lifted. The yoke supports the motor drive unit and the spindle assembly. On some shapers, the elevating mechanism raises only the spindle assembly (Illus. 2-11). The spindle assembly houses the spindle and the spindle bearings.

Illus. 2-11. On some shapers, the elevating mechanism raises only the spindle assembly. (Photo courtesy of Delta International Machinery Corporation)

Illus. 2-12. The lock knob in the center of the handwheel locks at any spindle height.

Illus. 2-13. This locking mechanism clamps the sleeve around the spindle. Any setting can be held securely.

Illus. 2-14. This scale helps you determine changes in the height of the cutter. It makes minor changes in cutter height easy to set.

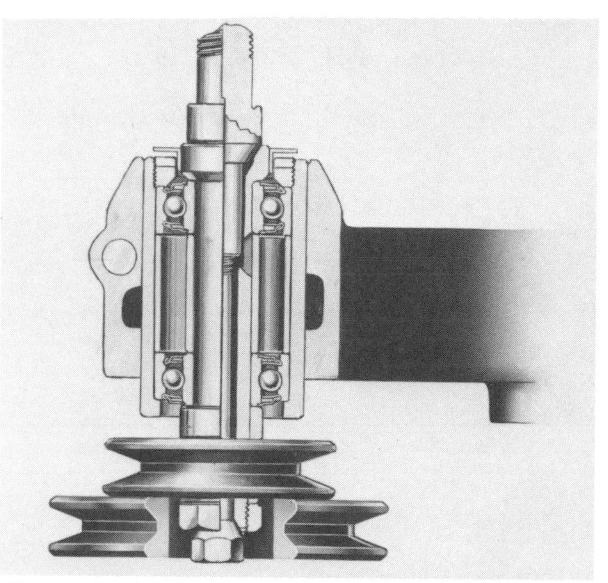

Illus. 2-15. The spindle assembly is the most important part of the shaper. It is loaded with ball bearings, and it drives the cutter. (Photo courtesy of Delta International Machinery Corporation)

The elevating handwheel has a locking mechanism to hold the desired setting. On some shapers, the locking mechanism is a knob in the center of the handwheel (Illus. 2-12). Other shapers have a lever located near the elevating handwheel (Illus. 2-13). It is turned clockwise to lock the spindle, and counterclockwise to release the lock.

The amount of spindle travel varies from one shaper to another. Spindle travel generally ranges from 2–6 inches. The greater the spindle travel, the more adjustment you have to make during shaping. Some shapers have an indicator to help you determine how far the spindle has been moved (Illus. 2-14).

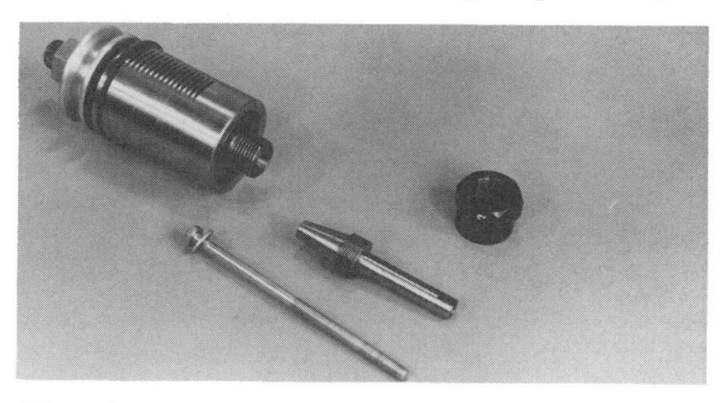

Illus. 2-16. Most shaper spindle assemblies have two sets of ball bearings: one at the top, and one at the bottom. The ball bearings help the spindle resist deflection under heavy shaping loads.

SPINDLES

The spindle assembly is the most important part of the shaper (Illus. 2-15). The assembly contains two sets of ball bearings (Illus. 2-16). The ball bearings resist deflection under heavy loads and ensure that the cut is smooth.

The spindle is what distinguishes a shaper from a router table. The actual spindle or arbor holds the shaper cutters. It is held in position by a drawbolt (Illus. 2-17). The spindle arbor is tapered for a perfect fit.

You can buy extra spindles as accessories for many shapers (Illus. 2-18). These spindles vary in diameter or function. Two common sizes are ¾ and 1 inch (Illus. 2-19).

Spindles can be as small as ½ inch in diameter and as large as 1¼ inches. Larger spindles can accommodate a stack of cutters and a heavy cut (Illus. 2-20). Smaller spindles cannot accommodate many cutters, and the depth of cut must be shallower. A heavy cut with a ½ inch spindle can cause deflection and poor quality cuts. It can also bend a small-diameter spindle. Do not overload the spindle.

Some of the specialized spindles available include stub spindles and router spindles. Stub spindles are sometimes used with cope cutters. You can use router spindles to adapt the shaper spindle into a router collet (Illus. 2-21). This means that you can use common router bits with the shaper. Router bits are considerably cheaper than shaper cutters that have the same profile. The versatility you gain by combining router bits and shaper cutters allows you to shape additional profiles.

Since most shapers have spindle speeds in the range of 7,000–10,000 rpm, the smaller-

Illus. 2-17. When interchangeable spindles or arbors are used in the spindle assembly, a drawbolt is used to pull the tapered part of the arbor into the spindle assembly. Here the drawbolt is shown on the left side. The drawbolt ensures a uniform fit between the arbor and spindle assembly.

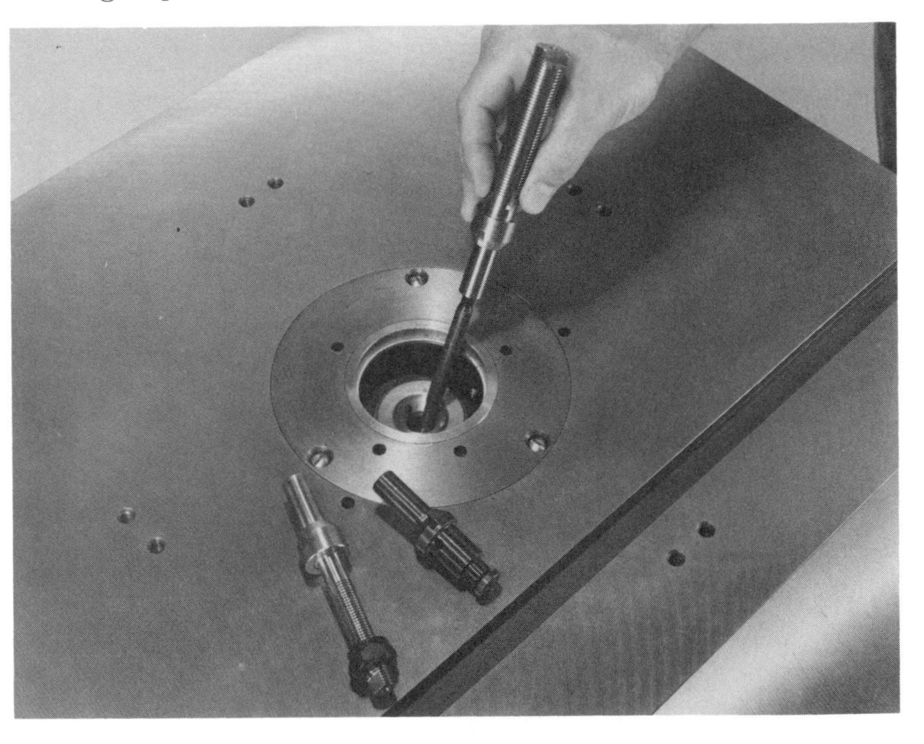

Illus. 2-18. Many manufacturers offer interchangeable spindles for their shapers. The diameters, heights, and functions of these spindles differ. In some cases, collets for router bits are also used to replace the shaper spindles. (Photo courtesy of Delta International Machinery Corporation)

Illus. 2-19. Spindles range from ½–1¼ inches in diameter. The spindle mounted here is ¾ inch in diameter. The other spindle is 1¼ inches in diameter.

diameter router bits are turning at about one-half their recommended speeds. Larger-diameter router bits have a greater peripheral speed and are not usually affected by the reduced spindle speed. Since the bits are turning at a reduced speed, you must also reduce the speed at which you feed the work into the cutter (Illus. 2-22).

If you plan to use router bits in the shaper collet spindle, use only those with a shank diameter of ½ inch or greater. Shanks with diameters smaller than ½ inch are not large enough to withstand the lateral stress generated when the workpiece is fed into the cutting portion of the bit.

It is also important that you consider spindle rotation when using router bits. When router bits are used in a shaper, the spindle should rotate counterclockwise. The stock should be fed into the bit from right to left, as you face the shaper. Before shaping with router bits, review Chapters 4 and 5 for information on safety and setup.

When using router bits in the shaper, make sure that the bit and the collet are compatible, that is, that the collet is clean and in good order. Check the bit to be sure that it is exactly ½ inch in diameter. If the bit is undersized, it could lift itself from the collet while it is under power.

Illus. 2-20. This spindle is 1¼ inches in diameter. The larger the spindle diameter, the larger the stack of cutters it can accommodate without deflection.

Illus. 2-21. If you use this router collet, you can adapt the shaper to use standard and custom-made router bits. Router bits are often much cheaper than shaper cutters.

Illus. 2-22. Router bits turn at a reduced peripheral speed. This means that the feed speed must also be reduced.

Illus. 2-23. Most interchangeable spindles have a tapered shank that fits into the spindle assembly. The taper helps ensure that the shank fits the mating portion of the spindle tightly.

This will damage your workpiece, and could cause an accident.

Changing a Spindle

Some shapers have interchangeable spindles. Shapers with interchangeable spindles offer greater versatility. They allow you to use cutters with different-sized spindle holes.

Most interchangeable spindles have a tapered shank that fits the mating portion of the spindle beneath the table (Illus. 2-23). You have to push or pull the tapered portion into a tight fit with the mating portion. This is accomplished with a drawbolt below the tapered portion and/or a nut above it (Illus. 2-24).

When changing a spindle, disconnect the power to the shaper. Clear away the fences and dust hood. Remove any cutters, shims, or collars before you begin.

On some machines the drawbolt under the spindle must be removed first (Illus. 2-25). On others, a nut on top of the spindle assembly must be removed (Illus. 2-26). In most cases, the spindle will be wedged tightly in place. In some cases, you can tap the drawbolt from the underside and drive the spindle up. On other machines, the spindle has a hole in it. You have to insert a soft metal rod into this hole. As the nut above the spindle is backed off, it bears against the rod and lifts it.

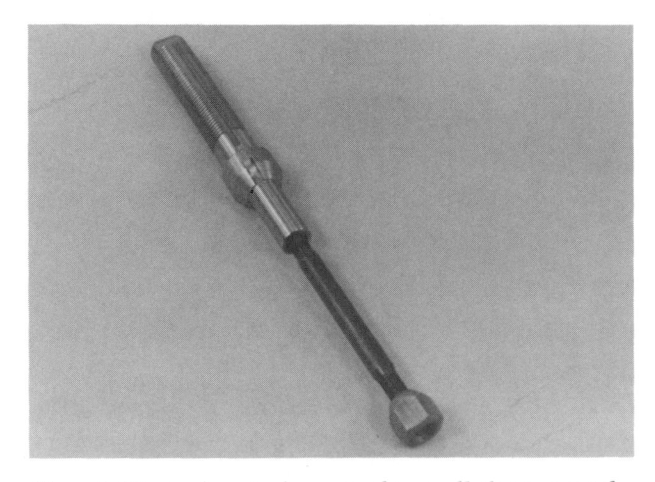

Illus. 2-24. A drawbolt is used to pull the tapered portion of the spindle into the spindle assembly.

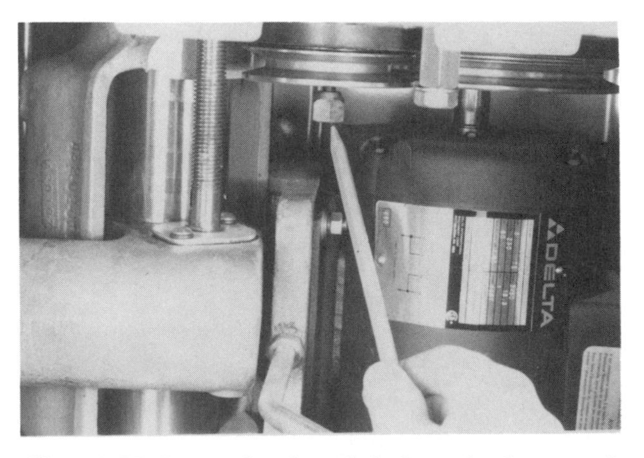

Illus. 2-25. Insert the drawbolt from the bottom of the spindle assembly, and loosen (or tighten) it with a wrench.

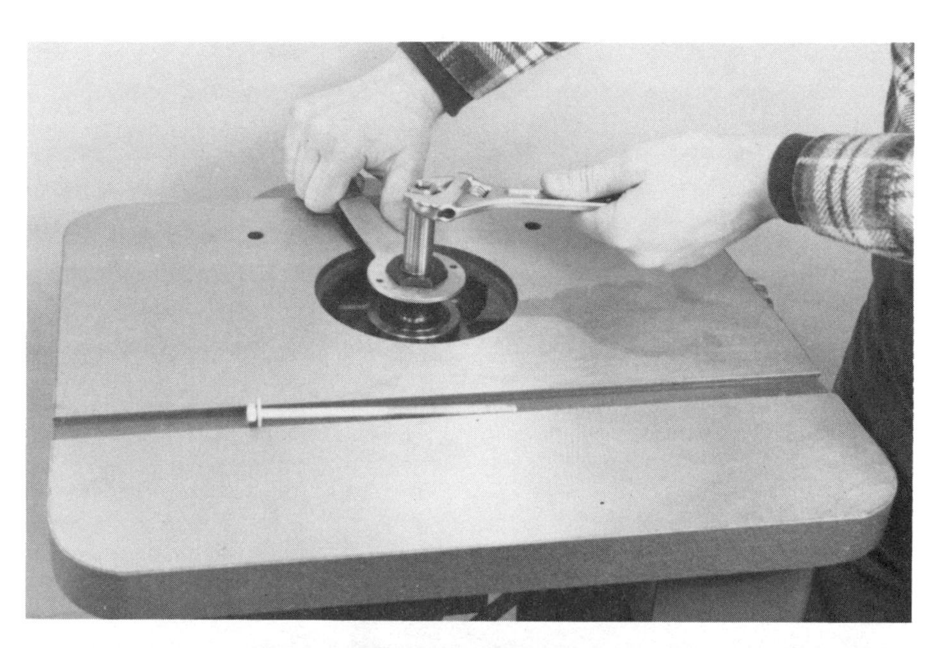

Illus. 2-26. When inserting a spindle on this shaper, you have to secure the spindle nut before tightening the drawbolt.

Illus. 2-27. When changing an arbor, make sure that the correct parts are all available before beginning.

Illus. 2-28. Note the pin on the spindle assembly. Some shapers have a pin to keep the spindle from slipping in the spindle assembly.

Illus. 2-29. Insert the spindle into the assembly. Make sure that the keyway is lined up with the pin.

Illus. 2-30. Secure the drawbolt to position the spindle or arbor.

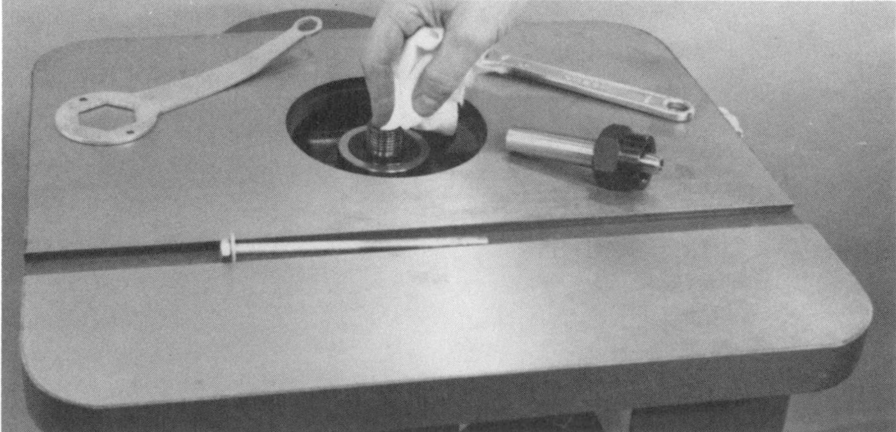

Illus. 2-31. After you have changed the arbors, wipe all the mating parts with an oily rag. This removes any dust or debris from the parts.

Illus. 2-32. This single-speed belt-drive shaper has a flat belt between the pulleys. The belt moves up and down on the pulley when the spindle is raised or lowered.

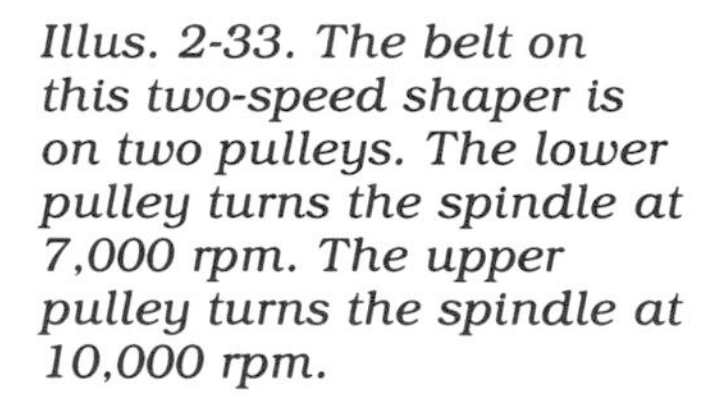

Illus. 2-33. The belt on this two-speed shaper is on two pulleys. The lower pulley turns the spindle at 7,000 rpm. The upper pulley turns the spindle at 10,000 rpm.

Illus. 2-34. If you want to change the speed of the spindle, you have to release the tension on the belt by using the thumbscrew (lower left) and the handle.

Regardless of how the spindle is changed, it is best to consult the owner's manual for your shaper. It will have specific instructions for spindle removal and replacement.

When you replace the arbor or spindle (Illus. 2-27–2-30), wipe the tapered portion with some light oil and a rag (Illus. 2-31). Do the same to the mating portion. This will remove any debris that may affect the fit between the two pieces. Tighten the spindle or arbor down securely, and turn it around by hand to be sure that it is seated properly.

Speed changes

Belt-drive shapers generally have a pulley on the spindle and motor (Illus. 2-32). On some shapers, the pulleys have more than one belt sheave (Illus. 2-33). To change speeds, you have to move the belt that goes between the pulleys to a different sheave. Make all speed changes with the power disconnected.

The position of the belt is usually changed through the access hole. In some cases, the belt is walked onto the next pulley. In other cases, the motor is released so that the belt can be repositioned (Illus. 2-34). Be sure to clamp the motor into position after you have adjusted the belt (Illus. 2-35). Adjust the belt tension according to the manufacturer's specifications (Illus. 2-36).

The spindle speed on direct-drive shapers is adjusted electrically. Usually, a speed-control

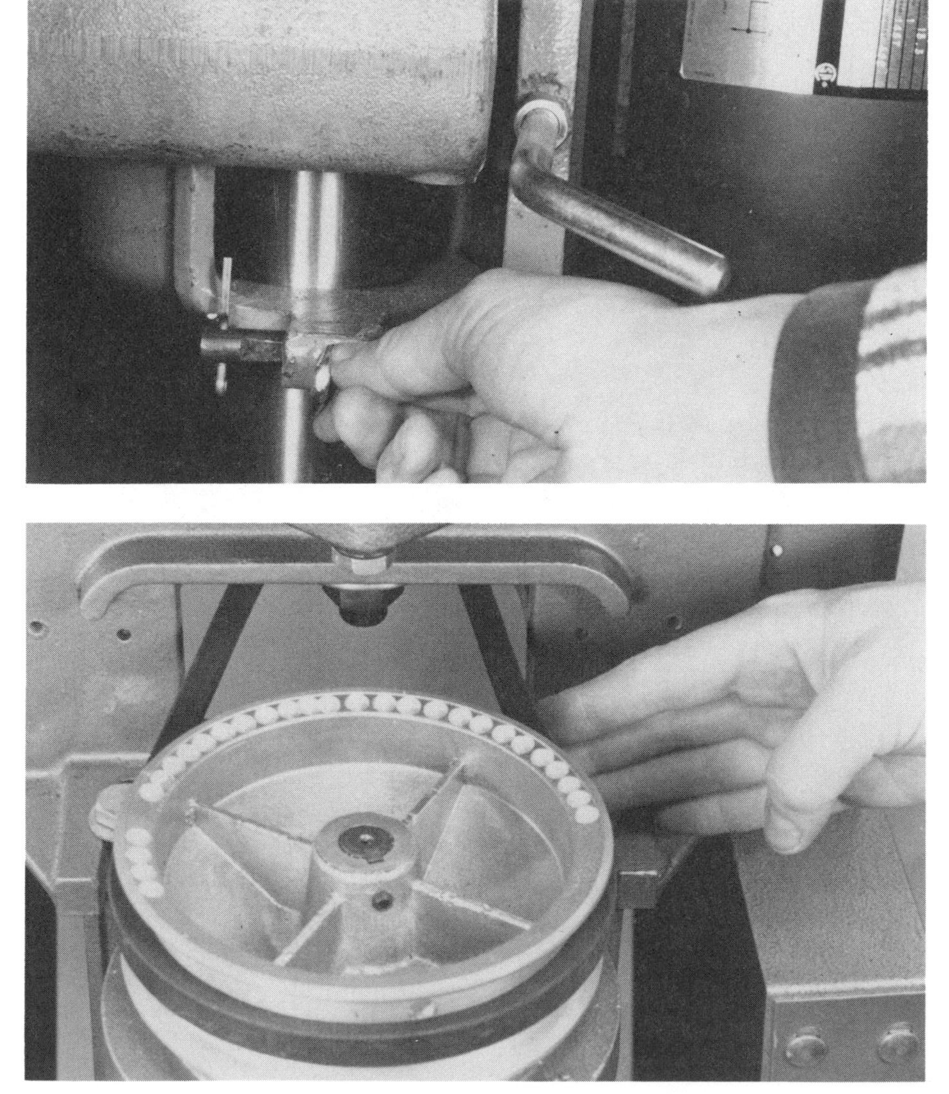

Illus. 2-35. Use this thumbscrew to lock the tension of the belt by which you have adjusted the speed of the spindle.

Illus. 2-36. Make sure that the belt on your shaper has been adjusted correctly. Check the owner's manual for specifics.

switch is positioned on the spindle near the on/off switch.

There may be a spindle-rotation switch, a speed switch, and an on/off switch on a direct-drive shaper. Make sure that you know the function of each switch before making any adjustments.

Spindle-tilting mechanism

A spindle-tilting mechanism is an option on some industrial shapers. To tilt the spindle, first release the locking mechanism. Turn the spindle-tilting handwheel until the spindle reaches the desired angle. A tilt scale is usually

Illus. 2-37. Angular fences can be used to tilt the work. This changes the profile made by the cutter.

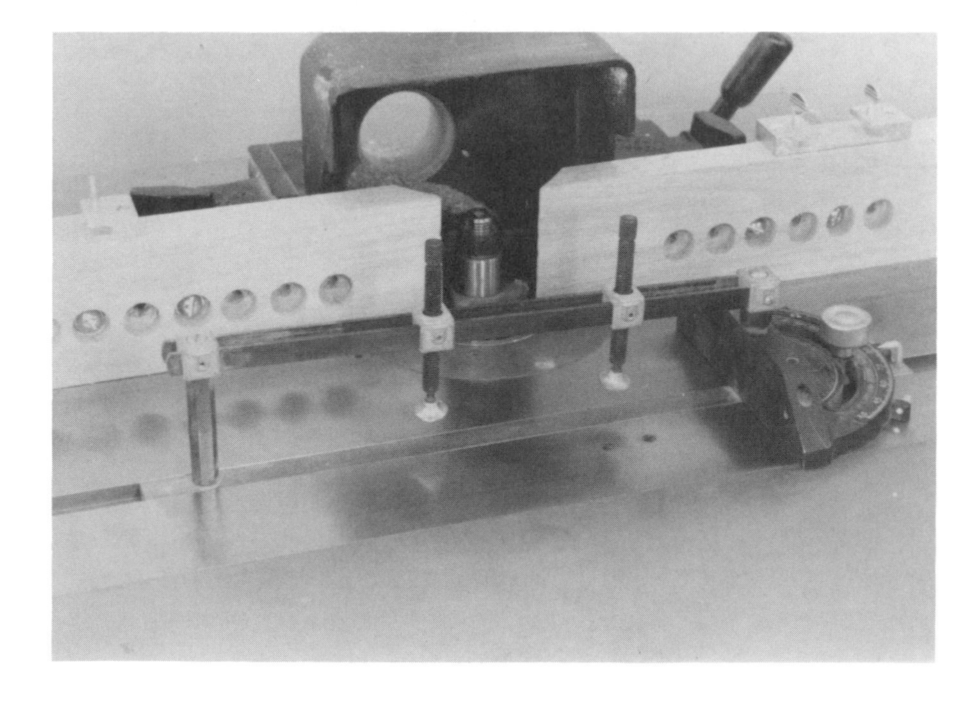

Illus. 2-38. The mitre gauge is an accessory that comes as standard equipment on all shapers. This mitre gauge has clamping mechanisms above the tongue.

provided to help determine the angle. Check this angle with a sliding T bevel or other layout tool to ensure accurate adjustment.

The chief advantage of a tilting-spindle shaper is that each cutter can cut different profiles when the spindle is tilted. This can reduce the number of cutters you need.

One of the most common complaints about tilting spindle shapers is that it may be difficult to adjust the spindle perpendicular to the table. If the spindle is not absolutely perpendicular to the table, mating cuts will not fit properly. If the spindle is tilted slightly, the tongue cutter will make the tongue slightly smaller than it should be. The groove cutter will then make the groove slightly larger than it should be. Always make sure that the spindle is perpendicular to the table before you set up the shaper.

If the spindle on your shaper does not tilt, you can use angular fences or tables to get the desired cut. You can either hand-feed or power-feed these fences (Illus. 2-37).

Illus. 2-39. This mitre gauge, shown on the right, has a T-shaped tongue. A wooden board has been attached to it. This helps eliminate grain tearout when you are shaping end grain.

Illus. 2-40. This mitre gauge has a dovetail-shaped profile on its tongue. It engages with the table and will not lift during shaping.

Accessories

MITRE GAUGE

The mitre gauge is an accessory that comes as standard equipment (Illus. 2-38). It fits into a slot in the shaper table.

On some shapers, the tongue of the mitre gauge engages with the table slot (Illus. 2-39). The profile of the tongue fits the slot profile. One common slot profile is the dovetail (Illus. 2-40), but others are also used. When the mitre gauge engages with the mitre slot, there is no chance that it will lift when the stock is being shaped.

The mitre gauge head can be turned for angular shaping. Loosen the lock knob and turn the head to the desired angle. Tighten the lock

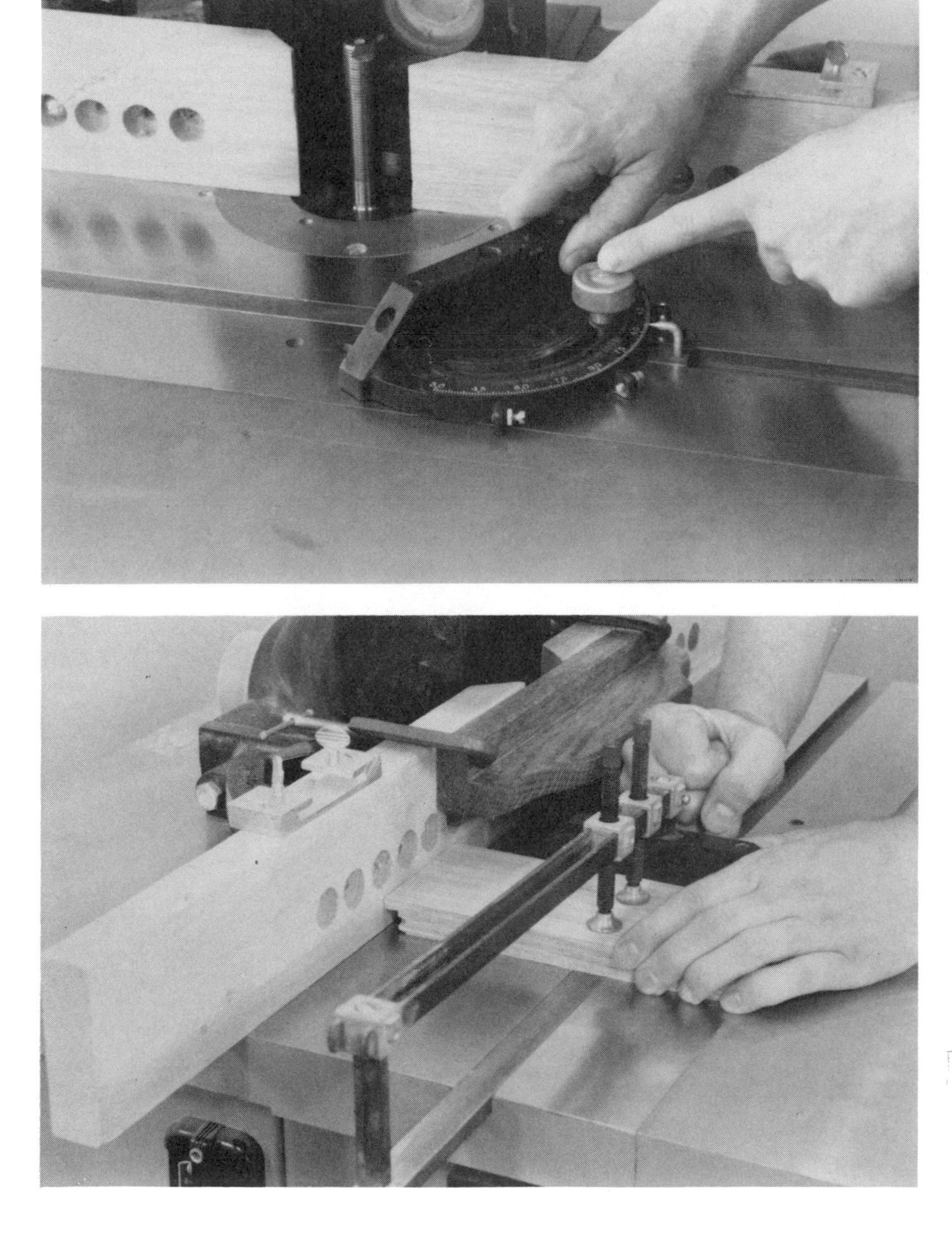

Illus. 2-41. To turn the mitre gauge, loosen the lock knob and turn the head to the desired angle. Lock the setting by tightening the lock knob.

Illus. 2-42. The clamping mechanism on a mitre gauge provides you with greater control when you are shaping end grain. End-grain parts tend to pull into the cutter while they are being shaped.

knob securely after making the adjustment (Illus. 2-41). The scale on the mitre gauge provides an indication of the angular adjustment. This is only an indication; check the setting with a layout tool for accurate adjustment.

The mitre gauge is used to control the stock when you are shaping end grain. Some mitre gauges have a clamping mechanism affixed to them (Illus. 2-42). This provides better control of the stock. The clamp keeps the stock from being pulled into the cutter as it is being shaped.

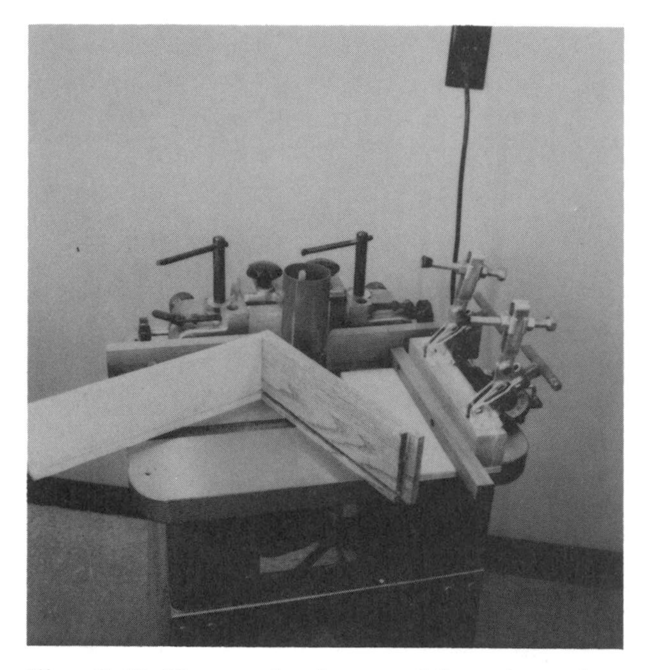

Illus. 2-43. The wooden base and face shown here were attached to the mitre gauge. You can use wooden bases and faces as a clamping mechanism when you are shaping end grain.

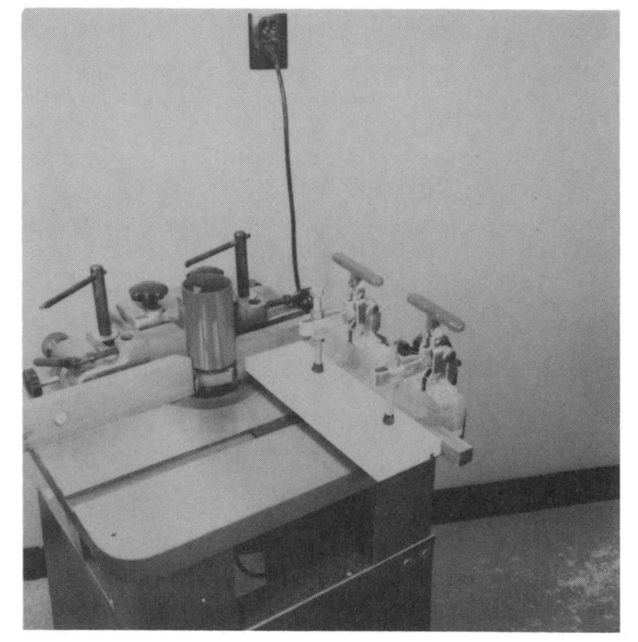

Illus. 2-44. These industrial clamps exert great clamping force on the work. They must be adjusted to the thickness of the stock you are shaping.

Illus. 2-45. You can clamp your own shop-made clamping mechanisms to the tongue of the mitre gauge. The industrial clamp exerts pressure right above the tongue of the mitre gauge.

Illus. 2-46. The thumbscrews are mating with holes that have been tapped in the tongue of the mitre gauge. The elongated holes in the wooden portion of the clamping mechanism allow it to be adjusted close to the workpiece.

On some mitre gauges, wooden faces and bases can be added for greater control of the work (Illus. 2-43). The industrial clamps lock down on the work (Illus. 2-44) and ensure that it does not creep or move during the shaping operation.

Specialty holding devices can also be attached to the tongue of the mitre gauge with threaded fasteners (Illus. 2-45). The industrial clamp holds the work between the mitre-gauge tongue and its contact pin (Illus. 2-46). Additional shop-made devices can also be added to make the job safer and more efficient (Illus. 2-47).

SLIDING SHAPER JIG

The sliding shaper jig is similar to the mitre gauge. It has a tongue that engages in the mitre slot.

Illus. 2-47. This hold-down has been positioned to exert pressure on the workpiece while the end grain is being shaped. The hold-down also acts as a barrier between the cutter and the operator.

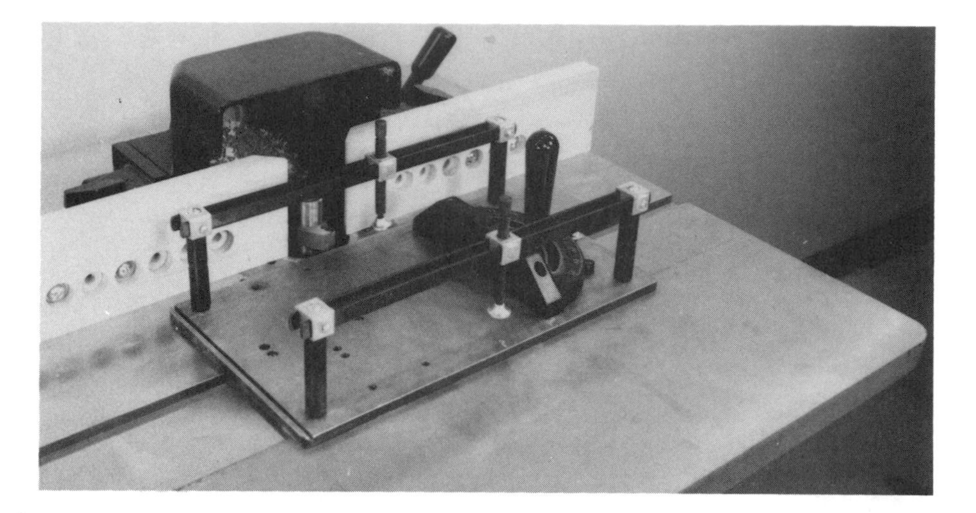

Illus. 2-48. The sliding shaper jig is similar to a mitre gauge. The tongue on this jig engages with the mitre slot.

Illus. 2-49. Work is clamped to the base with the overhead clamping mechanisms.

Illus. 2-50. This sliding shaper jig has a hand-clamping mechanism that holds stock securely for end-grain shaping. The hand-clamping mechanism adjusts to various thicknesses of stock. (Photo courtesy of L. A. Weaver)

The sliding shaper jig has a metal plate that rides on the table (Illus. 2-48). You clamp the work to the plate using the clamping mechanism that is suspended over the work (Illus. 2-49). Other sliding shaper jigs use hand pressure to secure the work (Illus. 2-50).

Some sliding shaper jigs have a head like the head on a mitre gauge. It can be adjusted for angular shaping. Adjust it the same way you would adjust the head on a mitre gauge (Illus. 2-51).

You can make sliding shaper jigs that are similar to the ones shown in Illus. 2-48–2-53. In most cases, shop-made jigs do not have an adjustable mitre-gauge head (Illus. 2-52). They are set for one angle only.

Some shop-made sliding jigs do not have a tongue that engages with the mitre slot. These jigs ride on the table against the fence (Illus. 2-53). The advantage of these types of sliding jig is that the fences do not have to be aligned with the mitre slot before use. This reduces setup time.

ROLLING TABLE

It is sometimes difficult to control large or heavy pieces when a mitre gauge or sliding shaper jig is being used. For this reason, some shaper manufacturers offer a rolling table. In some

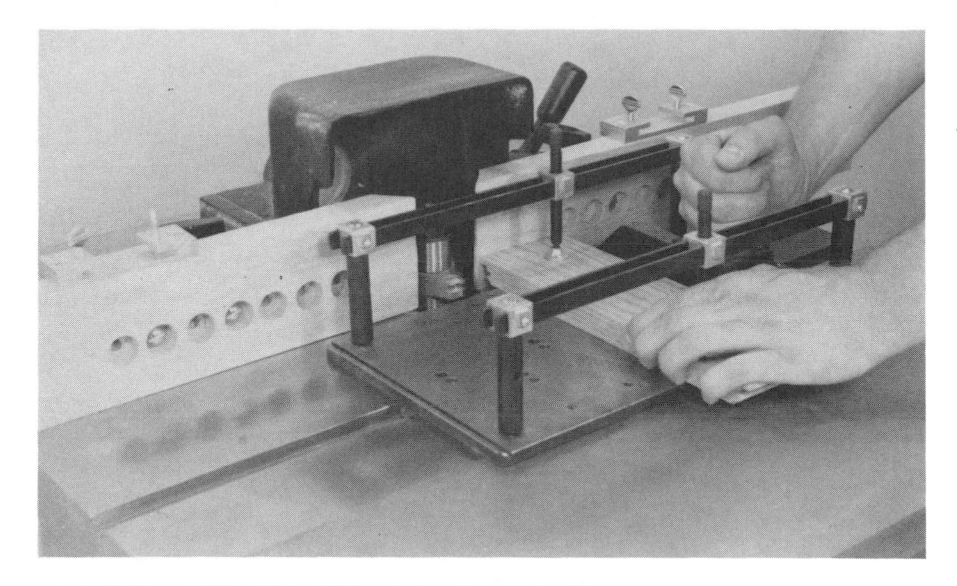

Illus. 2-51. The sliding shaper jig is adjusted in the same way as a mitre jig is adjusted. By turning the knob, you can turn the mitre head to any desired angle. Always lock the head securely after making the setting.

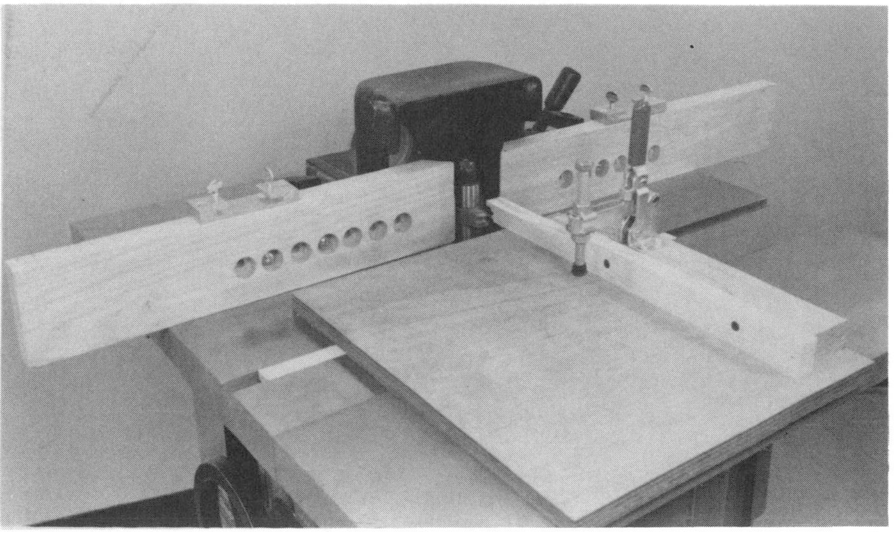

Illus. 2-52. This shop-made sliding jig is not adjustable. It is fixed at the desired angle.

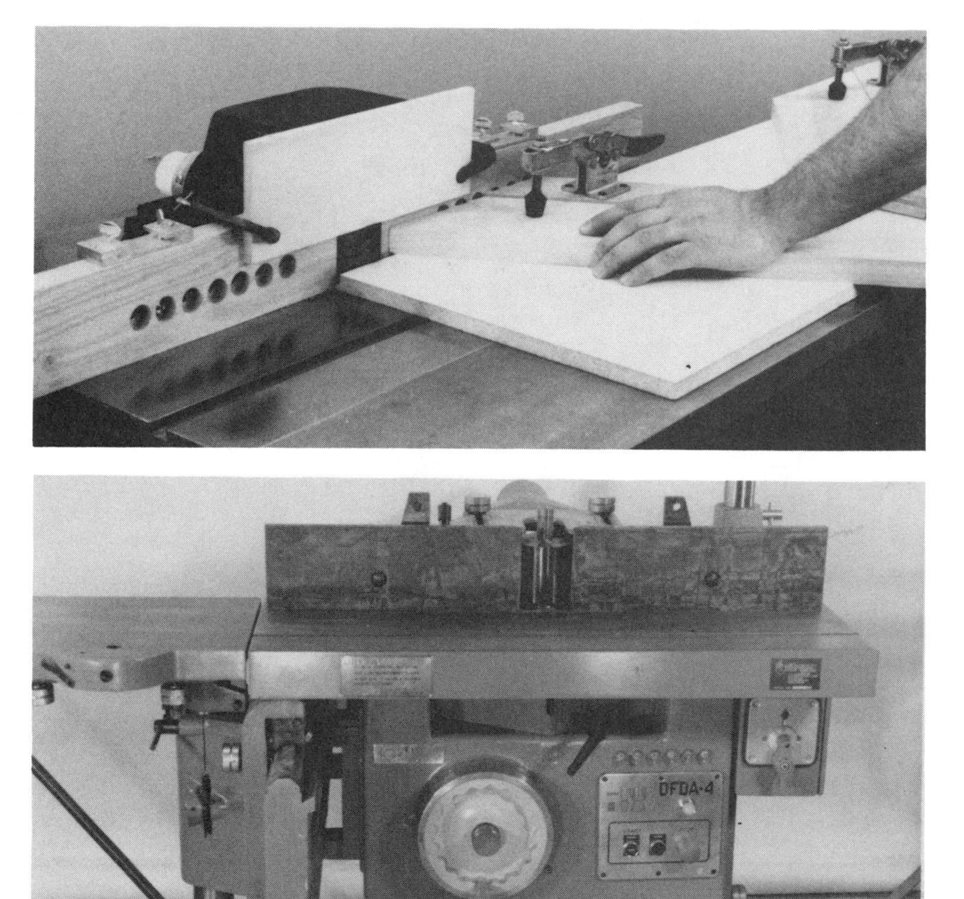

Illus. 2-53. You can make a sliding jig that does not have a tongue on it. This allows the jig to ride freely on the table and against the fence. You do not have to align the fences with the mitre slot when using this type of sliding jig.

Illus. 2-54. The rolling table on this shaper is part of its design. It acts like a take-off table in this position.

Illus. 2-55. Note the rail on which the rolling table slides. The table must be raised for use.

Illus. 2-56. When this handle is pushed down, it raises the sliding table above the actual shaper table.

Illus. 2-57. Once the table is raised, you can remove the handle.

Illus. 2-58. The fence assembly rotates on a turntable. It locks in a position perpendicular to the sliding table.

Illus. 2-59. The sliding table is now pushed towards the fence and locked at the desired position. Note: The rolling table must always be clear of the cutter.

Illus. 2-60. A clamping- and hold-down-type mitre gauge is mounted on this rolling table. It will control stock during the shaping process.

Illus. 2-61. This split-fence actually consists of two independent fences. These fences are adjusted separately.

cases, the rolling table is part of the shaper design (Illus. 2-54–2-60).

The rolling table supports heavy pieces. A mitre head is attached to the table for stock control. Sometimes there is also a clamping device, which also helps to control the stock. Heavy pieces are positioned against the mitre head and rolling table. The clamp holds the work securely. This enables the operator to concentrate on shaping the work. The weight of the work is supported by the rolling table.

Most rolling tables ride on bearings. It is important that the bearings and mating surfaces be kept clean. An accumulation of sawdust and wood chips can affect the smooth operation of the rolling table.

Wipe down the mating surfaces at the first sign of operating problems. Use a solvent such as mineral spirits to clean the parts. Do not use lubricating oils. These oils tend to attract dust and cause operating problems. Dry lubricants do not attract dust.

FENCES

All shapers offer a split fence as standard equipment. The split fence has separate fences (Illus. 2-61). Each side is adjusted independently.

Most of the controlling parts of the fence are made of metal (Illus. 2-62–2-64). The actual fences, however, are made of wood. They are attached to the metal portion of the fence with screws.

The wooden fences can be moved from side to side (Illus. 2-65). This allows them to be moved

Illus. 2-62. The studs attached to this table will secure the fence assembly to the table. They can be removed so that the table can be used for freehand shaping.

Illus. 2-63. The fence assembly goes over the studs. The threaded knob-type fasteners secure the fence assembly to the table.

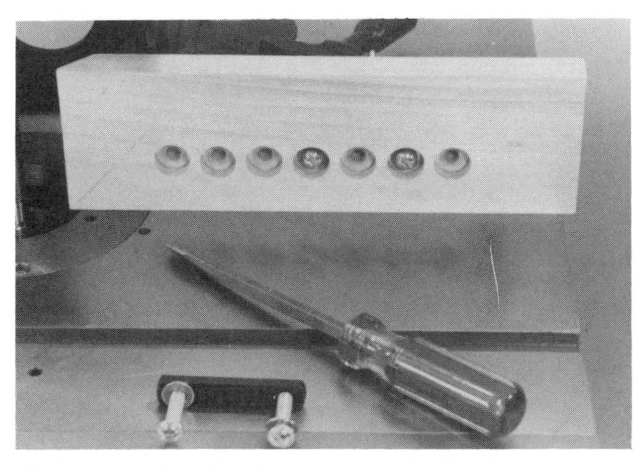

Illus. 2-64. Wooden fences are mounted to the metal fence assembly with the threaded fastening system shown here. The fences can be moved laterally towards (or away from) the cutter. The distance between the fence and cutter should always be minimal.

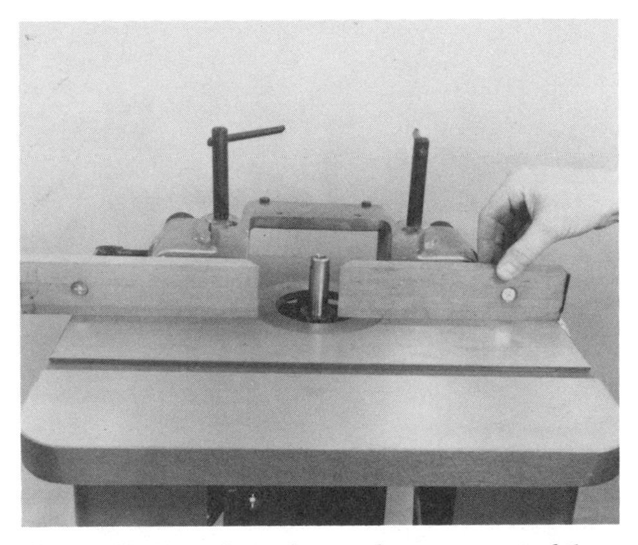

Illus. 2-65. Note the side-to-side movement of the wooden fences. They move independently of the metal fence assembly.

towards or away from the shaper cutter, depending on the diameter of the cutter.

The metal portion of each fence moves independently (Illus. 2-66 and 2-67). It can be moved in and out. This controls the depth of cut.

In some cases, you will be shaping away an entire edge. This requires that the infeed fence and outfeed fence be offset. You will have to move the outfeed fence closer to the front of the shaper. The amount of offset equals the amount of stock removed from the edge.

On many shapers, the split fence is part of a unit that also contains a dust-collection hood and provisions for hold-down attachments. The dust-collection unit provides a port for a dust-collection hose. It also houses the shaper bit and prevents it from making contact with the shaper cutter.

Dust collection is important during shaping. The removal of chips helps improve the quality of the cut. An accumulation of chips near the cutter can cause them to be pounded into the work by the shaper cutter. It is also possible to clog the cutter with chips if no provision for dust collection is made. Cutting the same chip several times will cause premature dulling of the cutter.

The hold-down provisions on the fence unit can make shaping safer and more accurate. Metal hold-down leaves must be adjusted to the size of the workpiece (Illus. 2-68 and 2-69). These hold-downs move with the fence, so they may require adjustment when you move the fence.

Shaper users often replace split fences with a single-piece wooden fence. The center of the fence is cut away to accommodate the desired shaper cutter (Illus. 2-70).

A single-piece fence can be made of sheet stock or solid wood. The materials selected should be hard enough to resist wear and clamping pressure. The clamps that hold the fence on the shaper can smash soft woods. When the wood is smashed, clamping pressure is reduced. This could cause the fence to give way during a shaping operation. It is important that you select the materials carefully.

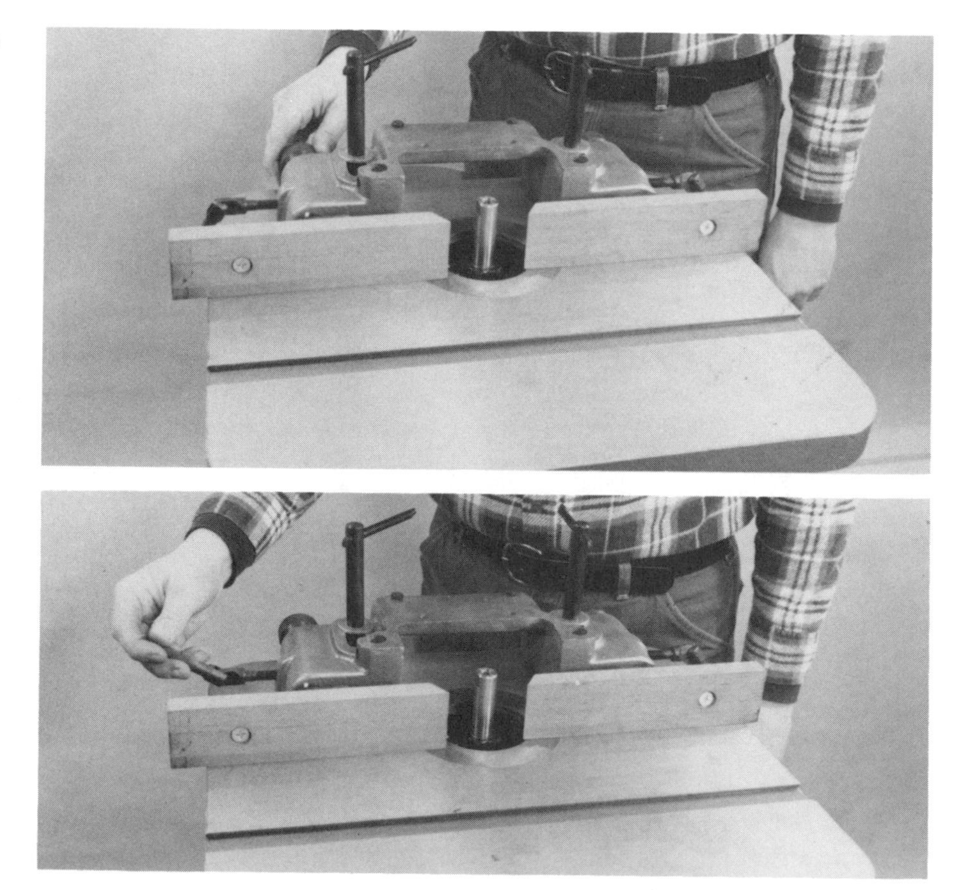

Illus. 2-66. The two fences on this split-fence system move independently. By moving the handle, you can move the fence towards (or away from) the mitre slot.

Illus. 2-67. The clamp handle locks the fence position relative to the mitre slot.

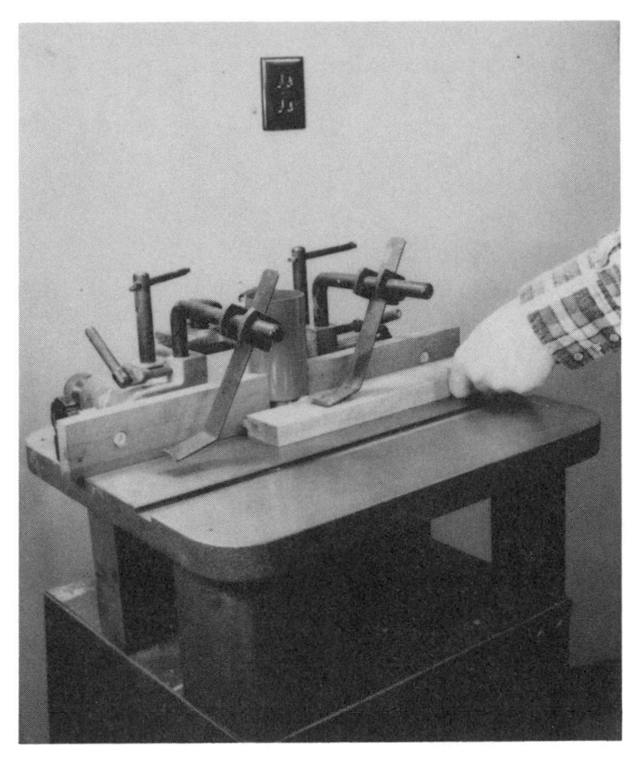

Illus. 2-68. These metal leaves hold the stock down against the table. They are secured to the fence assembly.

Illus. 2-69. The leaves on this shaper are secured to the table and fence assembly. (Photo courtesy of SCMI/Mini-Max)

Illus. 2-70. When no offset is required between the two fences, you can use a single wooden fence to replace them. Note how the fence has been coped (cut away) to accommodate the cutter and the arbor.

Illus. 2-71. When using split fences, keep the fences as close as possible to the cutter. These fences have been cut away to accommodate the guard.

Some shop-made wooden fences are designed to hold the workpiece at an angle. These fences change the way the workpiece contacts the cutter.

Shop-made fences can also be made to follow a circular profile on the workpiece. The fence is cut to a circular shape that complements the shape of the workpiece. These fences can only be used when the work has a circular shape. Irregular curves cannot be shaped with these fences. Irregular curves usually require a template.

When working with split fences or shop-made fences, it is important that you keep the fences as close as possible to the cutter (Illus. 2-71). The closer the fence is to the cutter, the safer the operation. A space between the fence and cutter can pull the work out of the operator's hands and into the opening. This can destroy the workpiece. It could also cause a kickback or bend the arbor.

For some shaping operations, it may be desirable to cope the shaper fences. Coping the fences means that the profile of the shaper cutter is sawn away from the fence. This allows the fence to be positioned very close to the cutter. There is little space between the fence and cutter. This keeps the workpiece from being pulled into the cutter.

One Delta shaper has a new, patented fence that can be used to universally adjust the cutter. The split fence is a series of V-shaped slats. These slats are made of a plastic material and are independently adjustable. They can be moved towards or away from the spindle. This allows a coping effect without cutting the fence. This shaper has other innovative features.

Cope the fences using a band saw, scroll saw, or coping saw. Use the cutter to lay out the fence. Outline the cutter with a pencil. Saw away the layout line. This will provide some clearance around the cutter when you position the fence. You may also have to cope the fence to accommodate the spindle nut and any part of the spindle that extends above the nut. The cope cuts will vary according to each individual setup.

GUARDS

Shaper guards prevent the operator from coming into contact with the cutter. There are numerous types of guards. Some attach to the fence to prevent the operator from contacting any portion of the cutter that extends beyond the fence. These guards may be made of wood or metal (Illus. 2-72).

Some guards surround the spindle. These guards are called ring guards because they are shaped like rings. The ring portion of the guard surrounds the spindle. It can be moved up and down to allow stock of various thicknesses to pass under it.

Illus. 2-72. This wooden barrier guard, which is clamped to the fence, makes it difficult for the operator to come into contact with the cutter during shaping. A shop-made device like this can be well worth the time spent making it.

Illus. 2-73. This ring-shaped guard is made of plastic and has a ball bearing in the center. It spins freely over the cutter when it is mounted on the arbor.

Illus. 2-74. This hold-down limits the operator's contact with the cutter. It has spring-tension adjustment that increases its hold-down strength.

Illus. 2-75. This ring guard mounts to the side of the table and limits your contact with the cutter when you are freehand shaping. (Photo courtesy of Delta International Machinery Corporation)

One ring-shaped guard attaches to the spindle (Illus. 2-73). It is a clear-plastic guard with a ball-bearing center. The guard spins freely over the cutter. Its diameter is greater than the diameter of all shaper cutters. This makes it almost impossible for the operator to come into contact with the cutter.

The edges of a plastic guard are smooth and free of sharp edges. If the operator comes into contact with this guard, he will not be cut (Illus. 2-71). This guard works well when you are shaping without fences. And it can be used with fences.

Some guards also act as a hold-down. They have a holding device attached to them. This provides positive holding force adjacent to the cutter (Illus. 2-74). These guards come in various configurations. Some are shop-made, and others are manufactured (Illus. 2-75).

Regardless of the shaping operation, a guard *must* be used. Stock should be held securely, and push sticks should be used to prevent the operator from coming into contact with the cutters. These other accessories are discussed in this chapter and in Chapter 4.

STARTING PIN

The starting pin is a smooth metal rod with threads or a taper on one end. The starting pin attaches to the shaper table in various places (Illus. 2-76). It is always positioned on the thrust side of the cutter. It supports irregular pieces as they are fed into a shaper cutter. The starting pin resists the thrust of the cutter as the shaping begins (Illus. 2-77).

Illus. 2-76. The starting pin is a smooth metal rod that attaches to the table in various positions. It should always be positioned on the thrust side of the cutter.

Illus. 2-77. The starting pin helps resist the thrust of the cutter when you start the cut. This gives you leverage and control over the workpiece.

Illus. 2-78. The starting pin is usually used when irregularly shaped objects are being shaped. Once the work touches the rub collar, pull it away from the starting pin.

Illus. 2-79. If a deep cut is required, take two or more light cuts. This reduces the chance of a kickback. Note how this guard can help limit contact with the cutter.

Starting pins are usually used when an irregularly shaped workpiece is being cut. A depth collar is used to control the cutting depth. No fence is used in this setup (Illus. 2-78).

The starting pin is positioned on the infeed side of the cutter. The work is butted to the starting pin before it contacts the cutter. The pin supports the work and provides leverage or a mechanical advantage over the turning cutter.

Do not attempt to shape irregular work without the aid of a starting pin. Take light cuts for greater control. Two light cuts are better than one heavy cut (Illus. 2-79).

Once the work is totally engaged with the cutter and depth collar, get away from the starting pin. Using the starting pin and the depth collar together could cause you to move away from the cutter. If you tried to go back into the cutter, there could be a kickback (the workpiece would fly at you).

Always set the work up so that you enter the workpiece on its edge grain. This way there is less tendency for the work to grab, and you can feed your work into the end grain. Freehand shaping is dangerous; always use a barrier guard and some type of jig to make the operation safer.

TABLE INSERTS

Table inserts are metal rings used to reduce the diameter of the spindle hole in the shaper table (Illus. 2-80 and 2-81). The hole in the shaper

Illus. 2-80. Table inserts are metal rings that go around the spindle. You can use them to adjust the diameter of the spindle hole according to the cutter being used.

Illus. 2-81. The holes in these table inserts have different diameters. Because table inserts have holes of varying diameters, you can select the insert designed for the cutter and operation being performed.

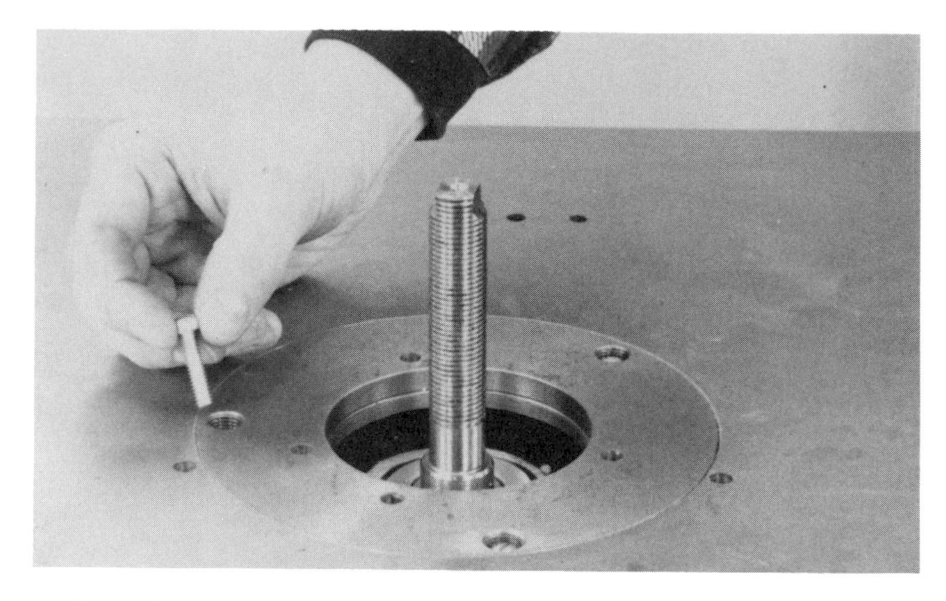

Illus. 2-82. This large table insert is held to the table with threaded fasteners. Always make sure that the table insert has been secured before starting any shaping operation.

table should be slightly larger than the cutter, unless the cutter is above the table (if the cutter is above the table, the cutter should be slightly larger than the hole in the table). This provides greater control over the workpiece.

A series of table inserts can be bought for most shapers. These inserts provide a close fit for a variety of cutters.

The table insert is usually held to the table with threaded fasteners (Illus. 2-82), a press fit (Illus. 2-83 and 2-84), or by gravity (Illus. 2-85). It is important that the insert is fastened securely in place, or that it is heavy enough to stay in place once you begin shaping. The turning cutter can sometimes develop enough momentum to raise the insert out of the table.

Some table inserts have threaded inserts in them which allow the table insert to be adjusted even with the shaper table (Illus. 2-86). This adjustment ensures uniform feeding and stock support. Check the alignment of the insert and the shaper table whenever you set up the shaper (Illus. 2-87).

COLLARS

Collars are metal rings that are mounted on to the spindle. There are three types of collars: spacer collars, shim collars, and rub collars.

Spacer collars are used between cutters to attain proper spacing. For example, a ¼-inch

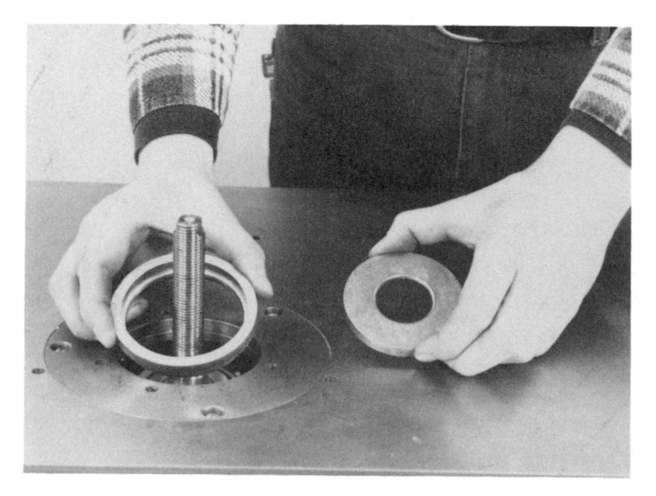

Illus. 2-83. These die-cast table inserts are compression-fitted into place.

Illus. 2-84. Note how the inserts have filled the hole around the arbor. You can use this setup if your cutter is positioned above the table.

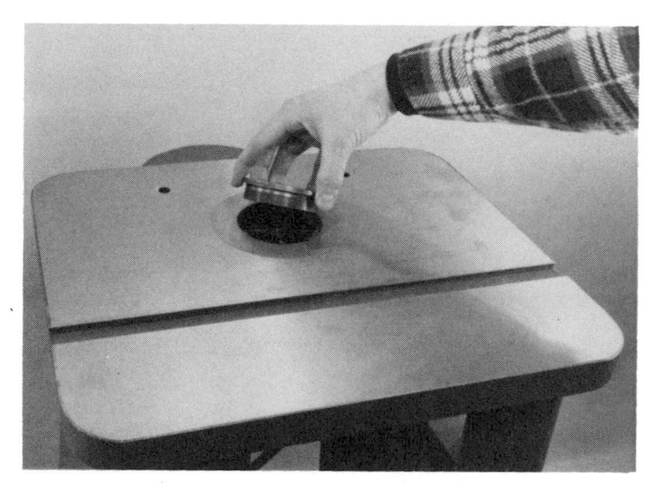

Illus. 2-85. These iron table inserts are held in position by their weight.

Illus. 2-86. The threads in this table insert allow it to be adjusted even, or flush, with the shaper table.

spacer collar would be used between two cutters when a ¼-inch tongue is needed between the cutters.

Spacer collars are available in various fractional and millimeter thicknesses. Some spacer collars are furnished with shaper cutters designed for joinery or door systems. They locate the position of various cutters in the set. Keep these spacer collars with the set at all times. They were machined for use with the set. Substituting spacer collars may adversely affect the fit of the mating parts.

Shim collars are also used to control the space between cutters. They compensate for the metal removed from the cutter when it is sharpened. These spacers are very thin. They are measured in thousandths of an inch.

Shim collars are frequently used with spacer collars. In some cases, they are added to a set of joinery cutters after they have been sharpened.

Rub collars are used to control the depth of cut. The rub collar rides on the workpiece edge and limits cutter depth. Rub collars can be positioned on the spindle over or under the cutter. The position depends on the operation.

Some rub collars are solid metal rings, while others have a ball bearing pressed into the center of a metal ring. Rub collars with a ball-bearing center turn freely when the work or template is pushed against them. This allows them to run cooler and more smoothly.

Solid metal collars do not spin freely. When they rub against the work or template, heat is generated. This heat can cause burning along the edge of the work or template. These burn marks are unsightly and difficult to remove. They also take stains (finishes) differently than the rest of the work.

EXTENSION WINGS

Extension wings are attached to the shaper table to extend the working area. Some extension wings are made of stamped steel, while others

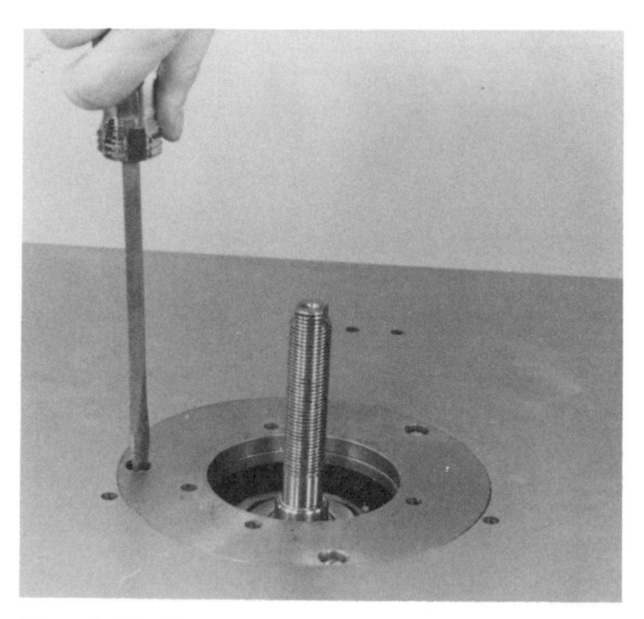

Illus. 2-87. When using this type of table insert, make sure before you proceed that it is even, or flush, with the table and that the fasteners are secure.

are cast iron. Cast-iron extension wings may be solid or ribbed. Most shaper wings can be bolted to the front of the table (Illus. 2-88), but they should be bolted to the infeed or outfeed side.

Extension wings provide greater control over large pieces. They help support the weight of the workpiece and help hold it in a true plane.

Extension wings are sold as an accessory for most shapers. Consult the manufacturer's accessory catalogue for details and specifications.

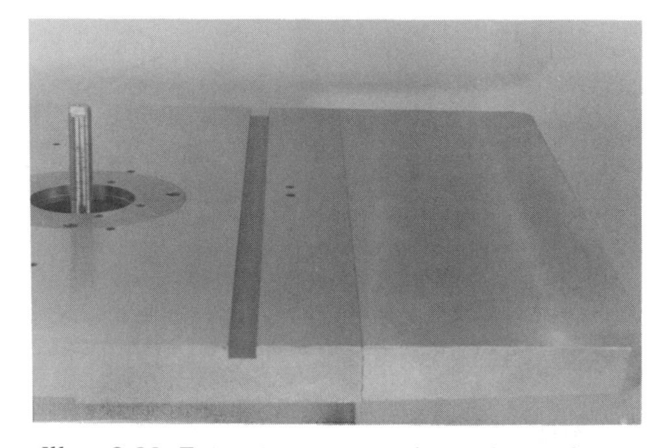

Illus. 2-88. Extension wings enlarge the working surface of a shaper. This is beneficial for some operations.

TAKE-OFF TABLES

Take-off tables are accessories used to support long pieces during the shaping process. Some

Illus. 2-89. The rolling table on this shaper doubles as a take-off table. It is locked in position for use.

Illus. 2-90. This shop-made extension table is clamped to the shaper table. It extends the working surface and supports long pieces.

take-off tables are connected to the shaper (Illus. 2-89), while others are positioned near the shaper when the need arises (Illus. 2-90).

Some tables have rollers on their surfaces. These rollers reduce friction between the work and the table. These roller tables must be positioned correctly for use. If the table is not level, the rollers could pull the workpiece from the desired path. Flat tables are less likely to do this.

In some cases, a dead man is used instead of a take-off table. A dead man is an adjustable stand that is positioned on the infeed and/or outfeed side of the shaper to support the work. It has a roller on top to allow the stock to move with little resistance. The dead man take-off is useful in a small shop. Again, the rollers can be a problem if they are not adjusted correctly.

You can make your own extension table. If doing so, make sure that it is made of stable wood or composite materials. Durable, high-density particleboard framed with hard maple would make an ideal extension table.

Always use good joinery techniques when making the table. Legs might also be helpful and are necessary if the extensions hang out more than 12 inches from the machine table. Plastic laminate is an excellent surface material for the tabletop. If you apply a good coat of paraffin to it, you will be able to move the workpiece freely and safely on it.

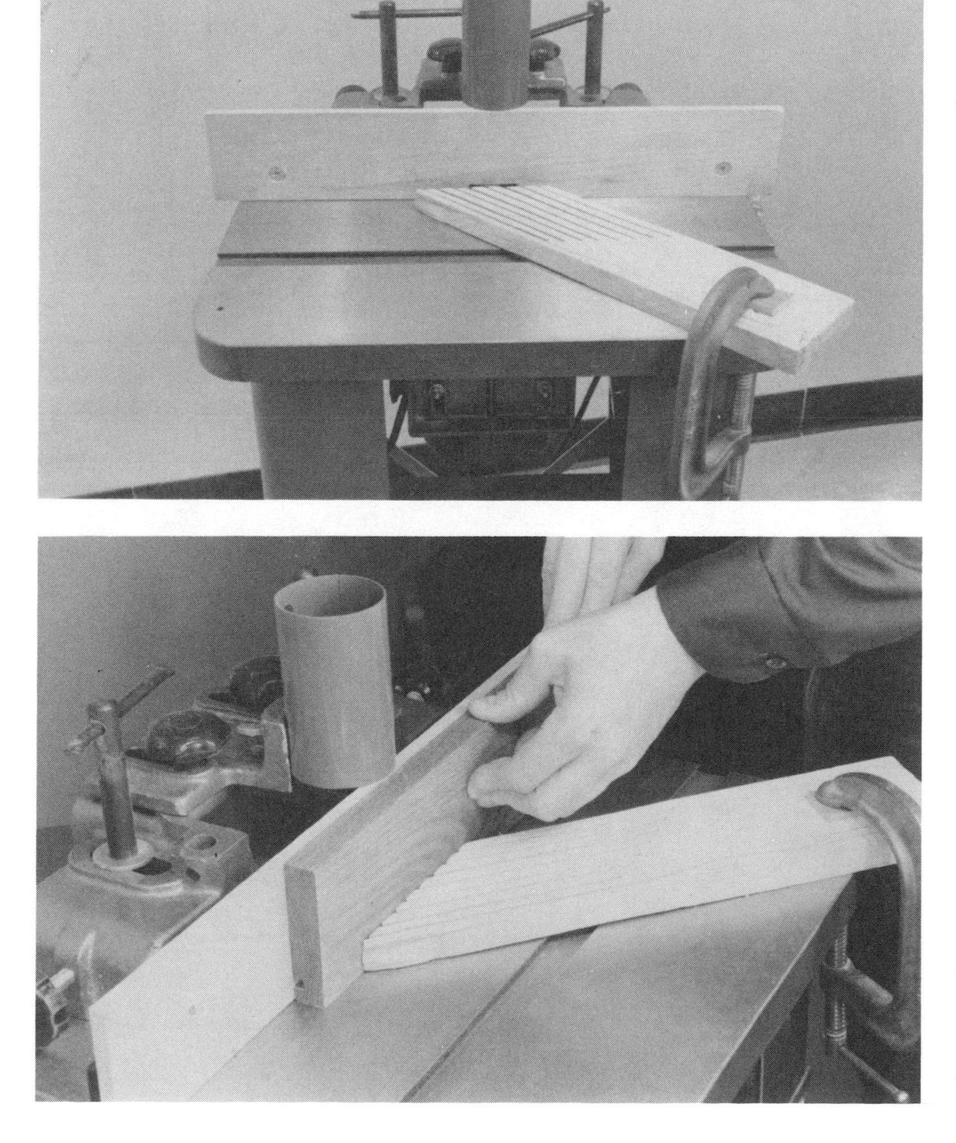

Illus. 2-91. This shop-made holding device is known as a featherboard. It holds stock against the fence or table.

Illus. 2-92. The feathers or tines actually deflect as the stock contacts them. This provides holding pressure and dampens vibration from the cutting.

You can also make extension tables that are freestanding. All four legs on a freestanding table should be completely adjustable. Once the table is adjusted for the operation at hand, you clamp or bolt it to the machine table to prevent any vibrations.

Keep the extension tables in good condition. This will help ensure better results when you are shaping.

HOLD-DOWNS

Hold-downs are wood, plastic or metal devices that are used to control the work. They are designed to hold the workpiece down on the table and/or hold the workpiece firmly against the fence. You can make your own hold-downs or buy them.

Shop-made hold-downs are referred to as featherboards or combs (Illus. 2-91). They are nothing more than a board with a series of cuts. The piece resembles a comb.

The featherboard hold-down is clamped to the fence or table. It is adjusted so that the wooden strips deflect as stock is fed into the shaper. This deflection acts as spring pressure to hold the stock against the fence or table (Illus. 2-92).

Manufactured hold-downs come either as metal springs or plastic wheels. Metal springs are usually attached to a metal rod. The metal rod is screwed to the fence or fence mechanism (Illus. 2-93). These metal hold-downs are adjusted so that they flex when stock is fed next to or under them. The spring action holds the stock correctly for shaping.

Shophelper® wheels are manufactured hold-downs that take the form of plastic wheels (Il-

Illus. 2-93. These manufactured hold-downs are actually hardened metal springs. Make sure that the hold-downs never contact the cutter.

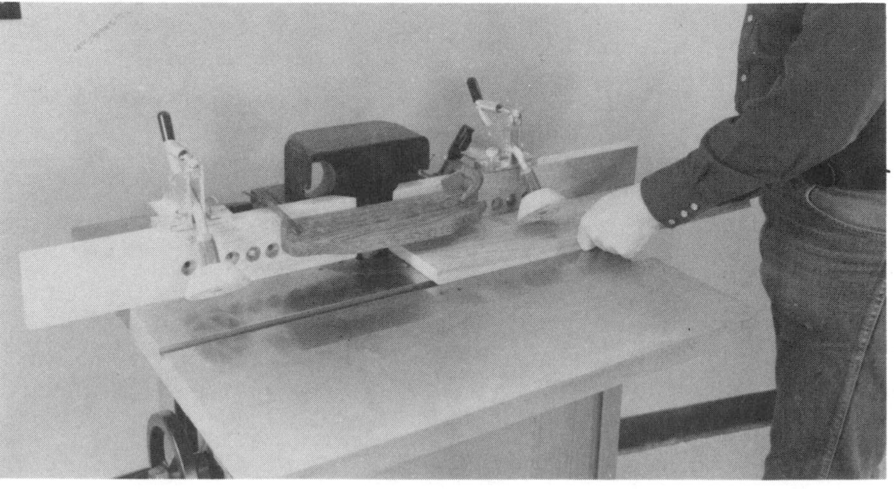

Illus. 2-94. These manufactured hold-downs are known as Shophelper wheels. The yellow wheels turn in one direction only. The green ones turn in both directions. Spring tension holds the stock against the fence and table.

lus. 2-94). These wheels are spring-loaded. They move in a track that is affixed to the fence.

Shophelper wheels turn only in the direction of the feed. This allows them to resist the force of a kickback. Yellow wheels are used for a right-to-left feed, and orange wheels are used for a left-to-right feed.

Shophelper wheels also hold the stock in towards the fence and down on the table. Their spring tension and height are controlled with threaded devices located behind the wheel.

Hold-downs of any kind make shaping safer. They reduce stock vibration and reduce the chance of kickback. Hold-downs can also act as a barrier guard that keeps your hands away from the cutterhead.

Use hold-downs whenever possible. Chapters 4–6 contain illustrations of different hold-down applications.

PUSH STICKS

Push sticks are shop-made devices. They are used to guide the work into the cutterhead. Push sticks take on many different shapes. The shape you select should be determined by the job you are doing.

Illus. 2-95. The power feeder automatically feeds stock into the cutter. It is suspended above the table and can be pivoted out of the way if necessary.

In some cases, two push sticks are used to control stock. This keeps both of the operator's hands well away from the cutterhead.

Study the push sticks shown in this chapter and throughout the book. Make several push

Illus. 2-96. The power feeder holds stock and feeds it into the cutter. This reduces vibration, wear on the cutter, and the chance of kickback.

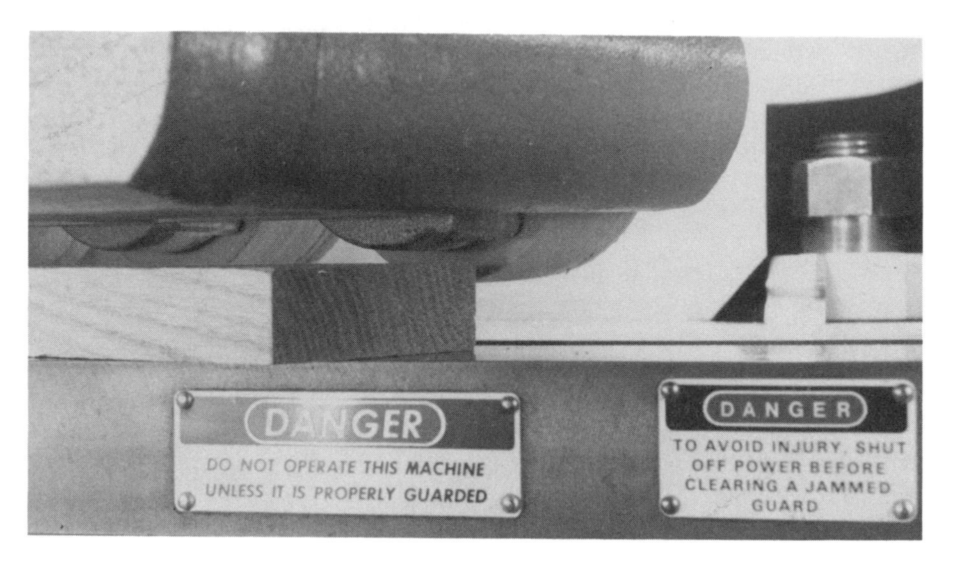

Illus. 2-97. The rollers on the power feeder hold the stock securely against the fence and table. To ensure that they have adequate traction, make sure that the wheels are about 1/4 inch lower than the top of the workpiece.

sticks before you do any shaping. Scale drawings appear in Chapter 4.

When making a push stick, remember to radius the edges of the handle to remove sharp corners. This makes the push stick more comfortable, and reduces the likelihood that the push stick will cut you in the event of a kickback.

Be prepared to make a new push stick when one is needed for a new job. Attempting to shape stock without the correct push stick is dangerous. The time spent making a new push stick is time well invested.

POWER FEEDERS

A power feeder is a device used to control stock during straight-line and some curved shaping. The power feeder consists of an electric motor and a series of feed rollers on a feed belt. The motor drives the feed rollers or belt, which, in turn, drive the work.

The power feeder is suspended above the shaper table (Illus. 2-95). It is suspended at a height that allows the feed rollers to grab the work and pinch it (Illus. 2-96). The pinching action feeds the work into the cutterhead. The wheels of the power feed unit must be about 1/4 inch lower than the top surface of the work you are feeding (Illus. 2-97). Consult the owner's manual for specifics.

Power feeders are sometimes set at a slight angle (Illus. 2-98). This forces the stock against the fence or table. Power feeders usually have an

Illus. 2-98. The power feeder is always oriented at a slight angle towards the fence or table. This ensures that stock will be held against both the table and fence.

Illus. 2-99. The access plate on this power feeder has been removed. You can change the speed of this power feeder by reversing the two gears closest to you.

adjustable feed speed. Speed is adjusted according to the speed and diameter of the cutterhead, the size of the cut, the hardness of the work, and the horsepower of the shaper. When in doubt, it is better to feed too slowly than too quickly. However, if you feed the work too slowly for the cutterhead, the work may burn along the cutting edge.

Power feeders feed stock at a uniform speed. This results in smoother cuts with less millmarks. The uniform feed speed also spreads wear over the cutters more evenly. This means that you will not have to sharpen the cutters as often. It also increases the amount of lumber shaped between sharpenings.

Probably the best feature of the power feeder is the fact that it keeps the operator's hands clear of the cutterhead at all times. The feeder controls the wood. The operator must only position stock under the power feeder and remove stock that has been shaped.

The power feeder also acts as a barrier guard. It also eliminates the possibility of kickback. The powerful motor ensures that the stock only travels in one direction.

While the power feeder makes shaping safer, there is a chance that your fingers can become pinched. Therefore, always keep your hands and fingers clear of the feed rollers. They could trap them between the work and table or between the work and feed roller.

Power feed units are used with standard split fences or custom-made or curved incline fences. Freehand shaping cannot be done with a power feed unit. Feed units are usually used to shape straight-profile work.

Power feed units come either with a belt or three, four, or six wheels. They have options such as forward and reverse feed, variable-feed speeds, or anywhere from two through eight feed speeds. On some units, you change the speeds by reversing the gears (Illus. 2-99).

The feed rollers are usually made of rubber or steel. Some feed units have a continuous rubber or plastic belt instead of rollers. Continuous rubber or plastic belts allow you to shape shorter pieces, since there is constant engagement between the feeder and work.

Feed units are available with single-phase or three-phase motors. Select the feed unit that matches your electrical system.

A power feeder will only feed effectively if its wheels are kept free of pitch and debris. If the wheels become scarred or cut up, they should be replaced. Check with your owner's manual or the correct replacement wheels.

PATTERNS AND TEMPLATES

Patterns and templates are devices used to control the stock while you shape. You can make your own or buy them (Illus. 2-100).

Patterns are usually designed to cut out an entire part. Templates are usually designed to

Illus. 2-100. Patterns and templates are used to control the work while you are shaping it. This pattern shapes the crown panel of a raised-panel door.

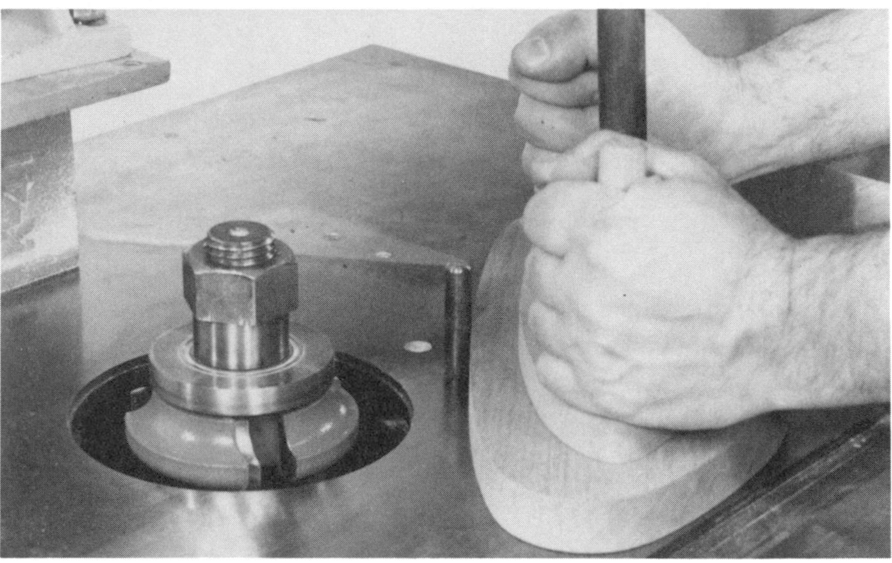

Illus. 2-101. A vacuum chuck holds stock securely. The vacuum will exert as much holding pressure as most clamps. The black vertical line shown here is the hose from the vacuum pump.

Illus. 2-102. This Panelcrafter is a manufactured pattern- and straight-work holding jig. It uses compressed air to hold the stock securely for clamping. Some air escapes from the underside of the jig. This makes movement of the jig much easier as it floats on the table.

Illus. 2-103. The Panelcrafter weighs about 70 pounds. The knob on the right handle actuates the air-clamping pistons.

cut out a portion of a part. Templates are sometimes used to cut out the arch on a crowned panel used in a raised-panel door.

Patterns and templates are usually controlled by a rub collar. The collar controls the depth of cut and the path of the cutter relative to the template or pattern. The work is secured to patterns and templates with mechanical clamps or metal fasteners.

In some cases, the stock is held to the pattern or template by a vacuum. This is known as vacuum-chucking. A vacuum chuck holds stock as securely as mechanical clamps (Illus. 2-101). Vacuum chucking is discussed in detail in Chapter 6.

Another type of pattern/template device uses compressed air to hold the stock securely for shaping. This device is known as the Panelcrafter™ (Illus. 2-102). In addition to clamping stock with compressed air, it allows some compressed air to escape through holes in the bottom of the jig. This escaping air allows the pattern/template to float on a cushion of air (Illus. 2-103). The result is decreased friction between the template and the table. This makes shaping easier and less fatiguing. The weight of the Panelcrafter™ also resists jolting or kicking caused by the cutterhead when certain parts are being cut. This makes the shaper less intimidating to the operator.

3
CUTTERS AND ATTACHMENTS

Shaper cutters (Illus. 3-1) are the key to making well-shaped profiles and tight-fitting joinery. The type of cutter you select will depend on a number of factors.

There are many different types of cutters. There are solid (Illus. 3-2) and assembled cutters (Illus. 3-3). The cutter may be made of high-speed steel or it may be tipped with Tantung™

Illus. 3-1. The type of cutter you select will depend on a number of factors. Read this chapter carefully to learn how to select the appropriate cutter. (Photo courtesy of Wisconsin Knife Works)

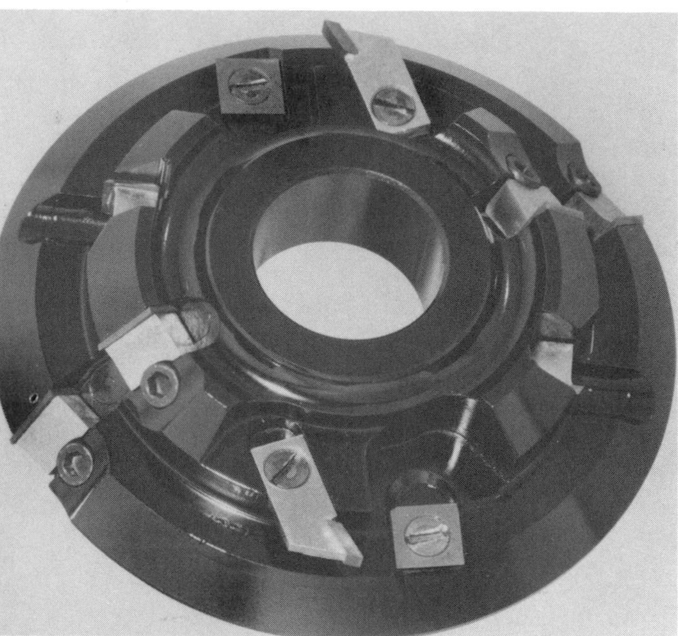

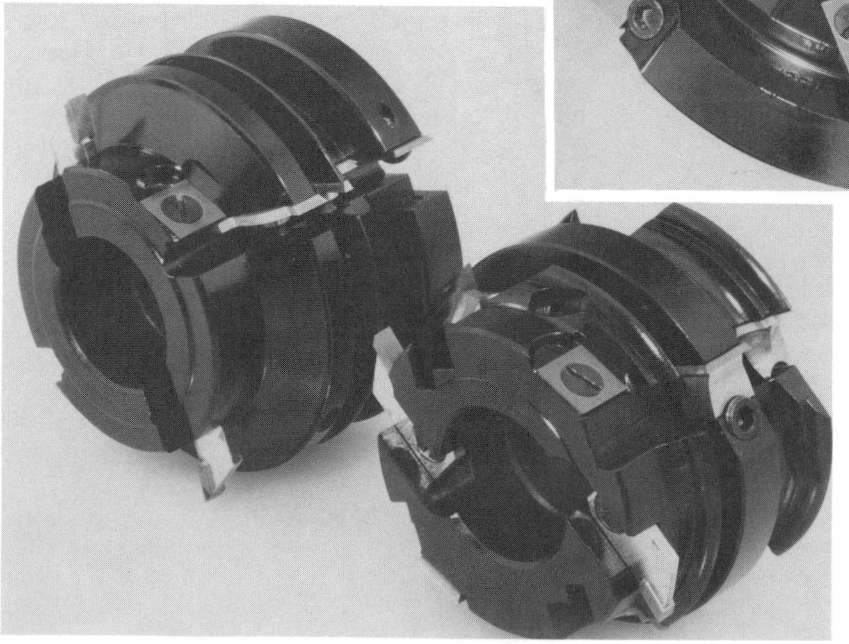

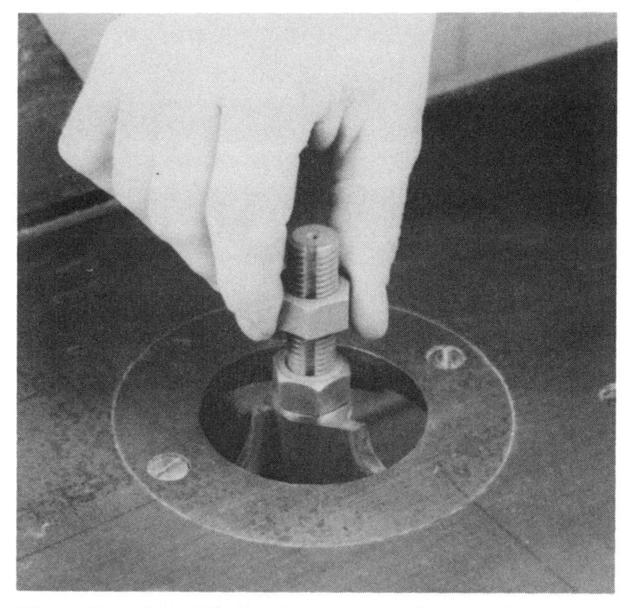

Illus. 3-2. A solid-steel cutter can be used to shape hard wood because it maintains a sharp edge.

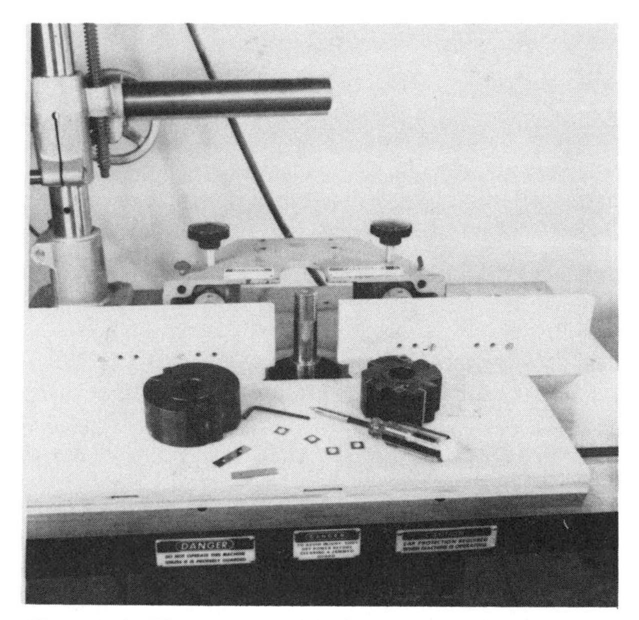

Illus. 3-3. These assembled cutterheads can be discarded when they become dull. This ensures that you will be using a sharp cutter at all times.

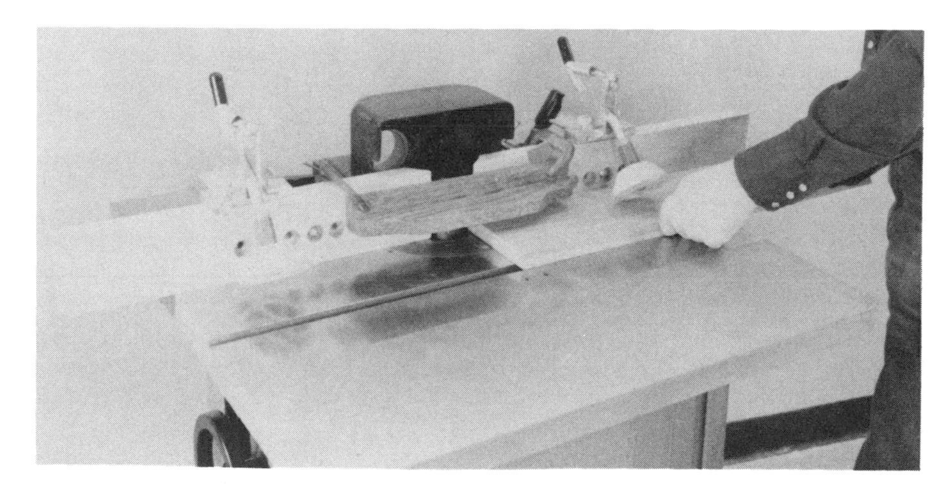

Illus. 3-4. Using the correct cutter for the particular job will make shaping safer and improve the quality of your work. Always check manufacturer's recommendations before selecting a cutter.

or carbide. The profile or shape of the cutter may be custom-ground or it may be a standard shape offered by many manufacturers.

This chapter contains all the information you need on which to base your selection. The correct cutter will make shaping safer and improve the quality of your work (Illus. 3-4).

Types of Cutter

The safest and most popular cutters used today for shaping are solid two- or three-wing cutters (Illus. 3-5). The number of wings denotes the number of cutting edges. These cutters are available from many different manufacturers.

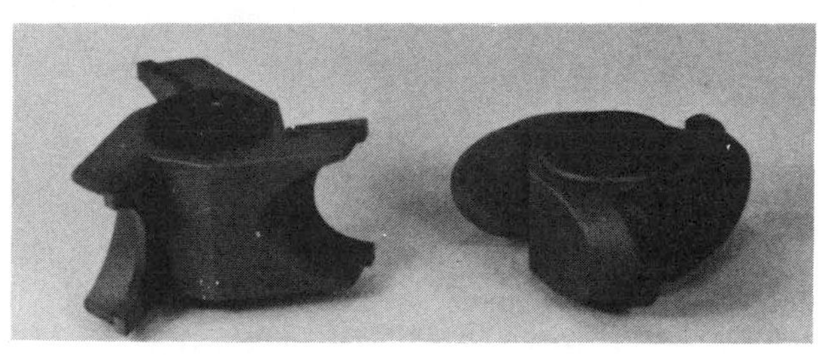

Illus. 3-5. Solid three-wing cutters are the most popular type of cutter used today.

Cutters vary according to size of the shaper's spindle (Illus. 3-6) and to the materials the cutting edge is made of. Solid shaper cutters can be entirely made of solid high-speed steel, or their tips can be made of tungsten carbide or Tantung (a cobalt alloy). These tips, when applied to the solid-steel blanks, are very hard, have great resistance to dulling, and can be used for a long time. Tantung cutters also have greater resistance to acid, so woods which have a high acid content will not break down the cutting edge.

Some manufacturers use copper between the carbide tips and steel body of a cutter to give it greater shock absorbency while it is cutting and limit vibration. A finer cut is achieved when vibration is reduced.

Two- and three-wing high-speed-steel cutters are designed with an involute relief angle, so they can be honed on their flat faces. This is advantageous because you can sharpen or hone them without changing their profiles (Illus. 3-7).

Solid cutters come in standard bore sizes: from ½–1¼ inches. The body width of the cutter depends on the profile of the cutter or the maximum thickness of the wood being cut.

Shaper cutters are generally made for either clockwise or counterclockwise rotation, though some cutters can be used for both rotations.

The rotation of the cutter determines which way the spindle will turn. It also dictates which side of the table will be used to feed stock into the cutter. Always consult your manufacturer's specifications to determine cutter rotation and the maximum rpm each cutter can be used at.

Illus. 3-6. The size of the shaper cutter varies according to the size of the spindle used on the shaper. Using cutters that are too large for your shaper could cause accidents and may damage the workpiece.

Illus. 3-7. Solid three-wing cutters usually have a flat cutting face. Therefore, you can hone them without changing their cutting angle. To sharpen them, hone the flat face only.

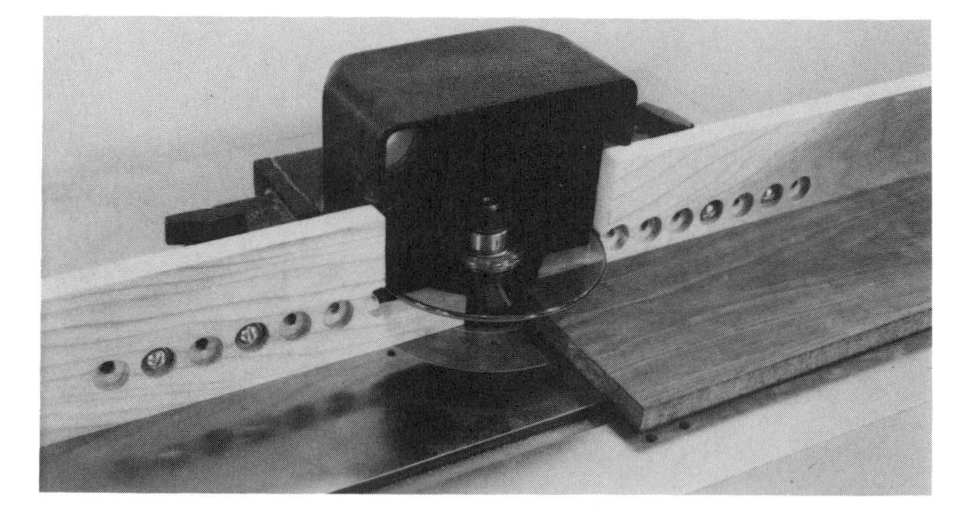

Illus. 3-8. The rotation of the cutter also determines whether the workpiece will be shaped with its good face up or down. A V cutter such as the one shown here allows the work to be shaped with its good face either up or down. This is the only cutter in common use that allows this.

Illus. 3-9. When the good face of the workpiece is up, the workpiece actually acts as a barrier between the operator and the cutter. This makes the operation safer.

Illus. 3-10. Raised panel shaping should be done with the appropriate jigs, hold-downs, and guards. In the setup shown here, there is a barrier guard over the raised panel cutter; also, a template rides along the pilot ball bearing to control the depth of cut. Your hands are kept clear, when you use the handles on the jig.

The rotation of the cutter will also determine whether the workpiece (stock) will be shaped with its good face up or down (Illus. 3-8). If its good face is down, this side will always be in contact with the table and could be scuffed or damaged during shaping. When the good face is up, the workpiece acts as a guard between the operator and the cutter (Illus. 3-9). This makes the operation safer.

Many raised-panel cutters are designed to shape the work with its good face up. This is not as safe, yet it keeps the panel from being damaged. Raised-panel shaping should be done with the appropriate jigs, hold-downs and/or guards (Illus. 3-10).

Some carbide and Tantung cutters are designed for specific spindle diameters. Larger-diameter cutters require spindles with large diameters (Illus. 3-11).

Most standard cutters are about 3 inches in diameter. They have a ¾-inch bore with a set of ½-inch bushings. The bushings allow the cutters to be used on a shaper with a ½-inch-diameter spindle (Illus. 3-12).

In the United States, the most-common spindle diameter is ¾ inch. This means that a cutter

Illus. 3-11. A large-diameter cutter such as the one shown here requires spindles that have diameters of 1¼ inches, and a shaper that has 3 horsepowers or more to power them. Smaller shapers cannot be used with these cutters. (Photo courtesy of Wisconsin Knife Works)

Illus. 3-12. The bushings shown with this raised-panel cutter allow it to be used on a shaper with a smaller spindle. These bushings are designed for a ½-inch-diameter spindle. The cutter is usually designed for a ¾-inch-diameter spindle.

with a ¾-inch bore is the most common and least expensive one available. For safety and quality of cut, use large-diameter cutters on shapers with 1- or 1½-inch-diameter spindles.

If you use cutters with carbide or Tantung tips, you will notice that the thickness of the tips will vary from one manufacturer to another. It is not always advisable to purchase a cutter with a thinner layer of carbide or Tantung, especially if the cutter you have purchased is a mating cutter. When a mating cutter becomes dull, the only way to sharpen it without changing its profile is to change or replace the cutting edges. It is more economical to change the cutting edge on a carbide or Tantung cutter. If this carbide or Tantung bit is a mating cutter, it is more economical to replace the cutting edge with a thinner rather than thicker layer of carbide or Tantung. For most applications, 3⁄32-inch-thick carbide or Tantung is appropriate.

Two- or three-wing solid cutters are standard for most shaping operations. There is a certain amount of runout that occurs when a three-wing cutter is being used, and, as a consequence, only one of the three knives actually makes the cut. This can cause chatter and premature dulling. The two-wing cutter has less resistance when it cuts. This results in less heat buildup, more even wear on the cutter, and, ultimately, a cleaner cut.

Tantung, a cobalt alloy, can be used in place of carbide for most applications. It produces a keener edge and can be honed easily, when necessary. It is also corrosive-resistant to the acids in wood. Additionally, it shears wood fibres cleanly and reduces the amount of sanding needed. Tantung cannot be used on remanufactured wood products such as particleboard or fibre-core plywood. One cut in remanufactured wood will dull the Tantung cutter. These cutters are usually painted yellow so that they will not be confused with carbide cutters.

ASSEMBLED CUTTERS

Cutterheads used to hold interchangeable profile cutters are made of steel or aluminum. The profile knives are held in a cutterhead in one of

Illus. 3-13. Various cutters that can be used with this interchangeable cutterhead. This cutterhead has a locating pin-and-socket system that holds the cutter securely during shaping.

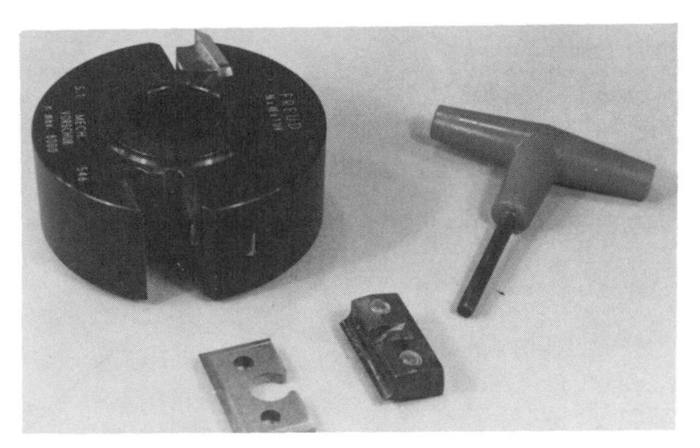

Illus. 3-14. A special wrench is used to remove and replace the cutters on this cutterhead. Always make sure when installing cutters in this cutterhead that the cutters are identical.

Illus. 3-15. The cutters on this assembled cutterhead are held in place with a gib screw and gib system. Special wrenches and screwdrivers are used to secure them properly.

Illus. 3-16. The gibs used in this cutterhead have special pins on them that mate with the cutter. This ensures that they are locked securely in position for shaping.

Illus. 3-17. All cutters used in this cutterhead are disposable. As the cutters become dull, they are flipped over so that their opposite side can be used. When this side becomes dull, the cutters are discarded. The square cutters that cut the shoulder of a rabbet have four cutting edges. They are indexed 90 degrees for each cutting edge.

Illus. 3-18. When using any type of cutterhead in which the cutters are inserted, make sure that the cutters are securely fastened. Check them periodically during operation to be sure that none have worked their way loose.

Illus. 3-19. This gib screw and gib system hold the cutters securely in place for shaping. Note the various profiles that are available for use with the cutterhead. This reduces the cost of buying cutters significantly. (Photo courtesy of Wisconsin Knife Works)

Illus. 3-20. This cutterhead has a gib screw and gib system. The cutters used in this system have two wings and chip limiters. The chip limiters control the depth of cut. They are discussed later in this chapter.

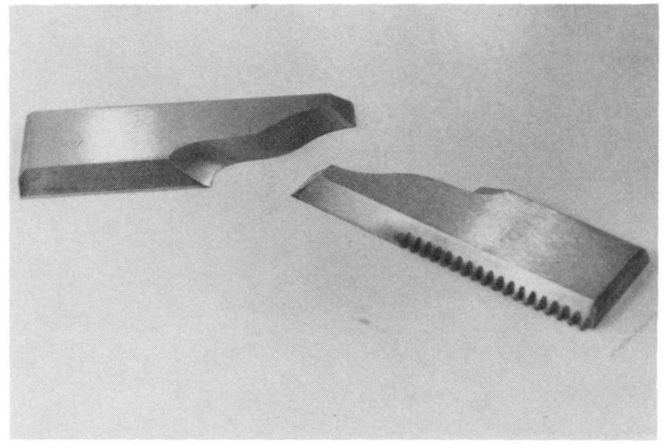

Illus. 3-21. These notch-edge cutters are inserted into a cutterhead which has mating grooves to accommodate the notches. These cutters must be set correctly for safe use.

several ways: a locating pin-and-socket system (Illus. 3-13 and 3-14); a gib screw and gib (Illus. 3-15–3-20); a notched edge used in a safety lock; and pinned collars (Illus. 3-21–3-23).

One cutterhead can be used with a number of different profiles (Illus. 3-24–3-27). Collars have proven to be a very economical way to use many different profiles on one cutterhead.

The knives used in the assembled cutterhead are made either of solid carbide or M-2-tempered high-speed steel. Some sets of cutters come with untempered blanks (Illus. 3-28) that can be ground to any desired profile, and then tempered prior to use.

Cutter knives are sometimes reversible; that is, they can run in either direction. Sometimes the cutterhead is turned upside down; in other cases, the knives are turned upside down. Always consult the manufacturer's directions when using any cutterhead.

Cutterheads usually have two knife inserts, but they can also have many more depending on the profile, diameter of the cutterhead, and the manufacturer's recommendations. Some cutterheads come with a rub collar that is part of the cutterhead assembly. This allows the cutterhead to be used for template and freehand shaping. The rub collar controls the path of the work or template (Illus. 3-29).

The diameter of the cutterhead assembly should match the size of the profile knife being used. For instance, a raised-panel cutter requires a larger cutterhead diameter than a radiusing cutter. Also, the larger the cutterhead, the larger the bore size of the cutter should be. This reduces chatter and deflection when you are making a large cut.

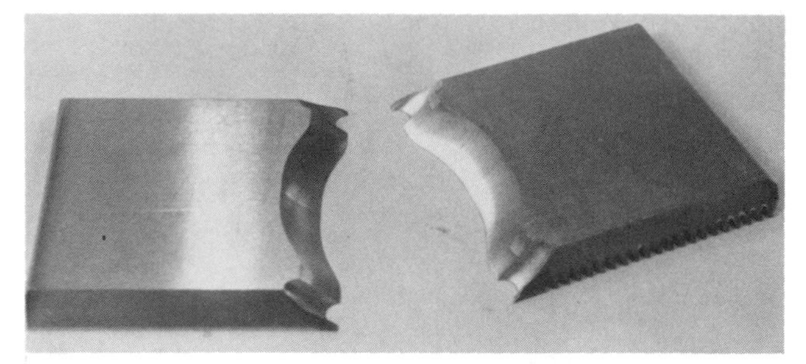

Illus. 3-22. The profiles and sizes of notch-edge cutters are limited only by the size of the cutterhead. Custom profiles can be easily ground in the high-speed steel-cutters shown here.

Cutterhead revolutions per minute (rpm) are prescribed by the manufacturer along with the specifications on the size of the knives to be used in the head assembly. Cutterhead assemblies usually should not have a bore less than ¾ inch in diameter, and more than 1¼ inches. The profile depth and width for each cutterhead assembly is also specified by each manufacturer. Check all specifications before beginning any setup. Never exceed the manufacturer's recommendations.

Illus. 3-23. When setting up notch-edge cutters, use a dial indicator to ensure that both cutters are turning in the same circle. This ensures that both cutters are removing wood uniformly.

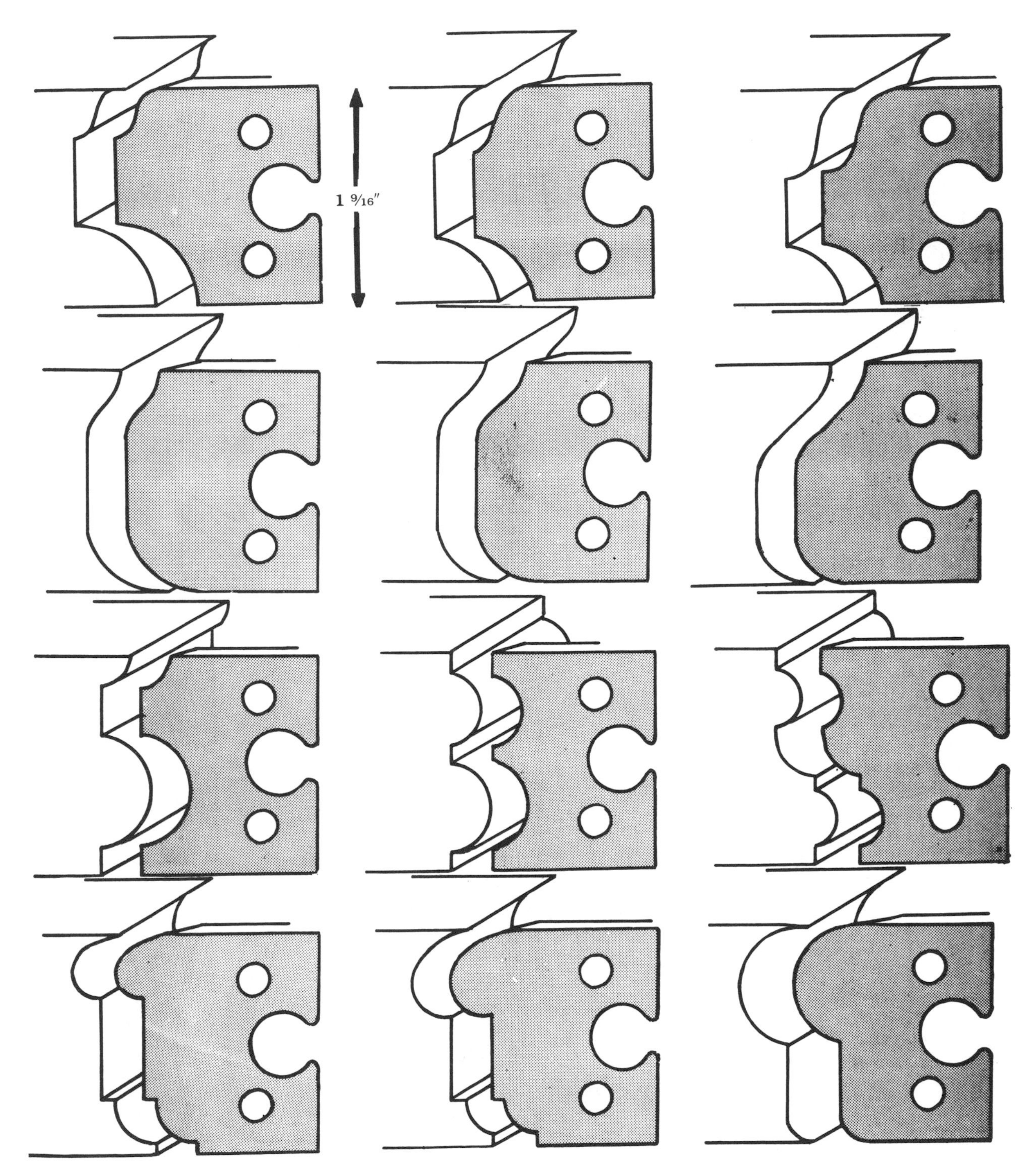

Illus. 3-24. These types of cutter can be used in an assembled cutterhead. Note the various shapes that can be made with these cutters. (Drawing courtesy of L. A. Weaver Co., Inc.)

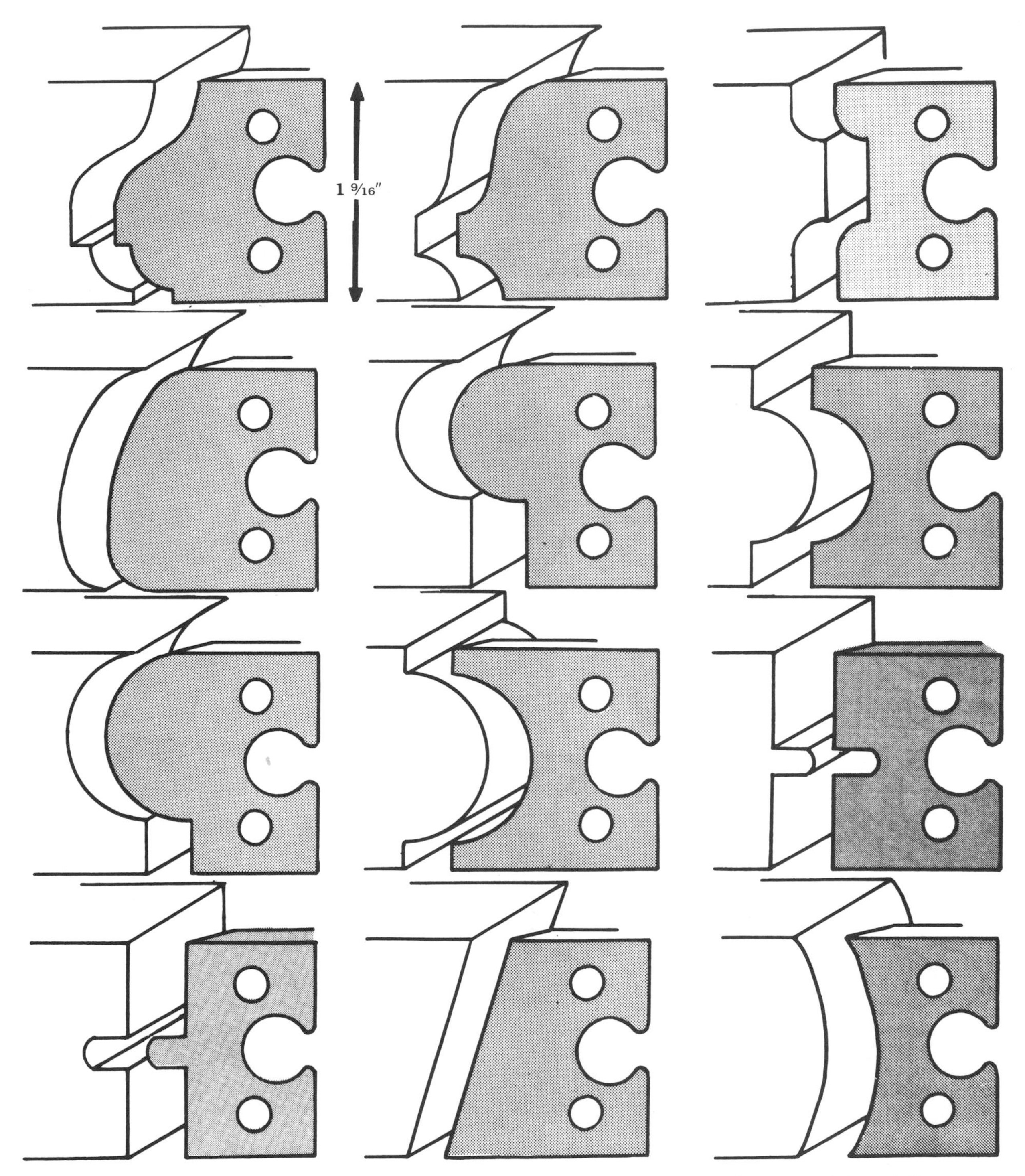

Illus. 3-25. These cutters can also be used in assembled cutterheads. All of the cutters shown here are made of tool steel. They are not designed to be used on plywood or particleboard.(Drawing courtesy of L. A. Weaver Co., Inc.)

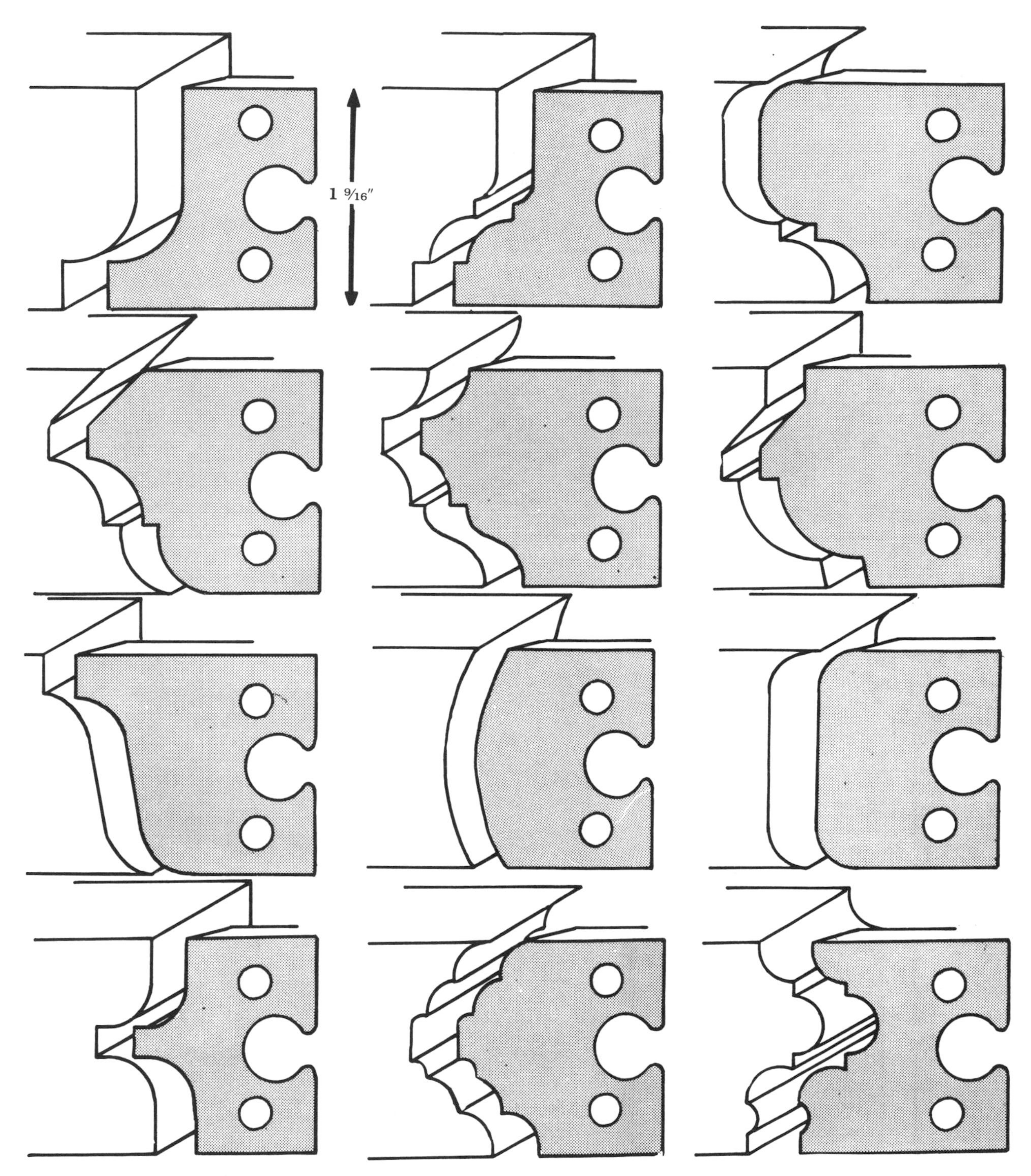

Illus. 3-26. These cutters can also be used in the assembled cutterhead. The holes in these cutters align with pins in the cutterhead to ensure safety during shaping. Always be sure that they are tightened securely before working. (Drawing courtesy of L. A. Weaver Co., Inc.)

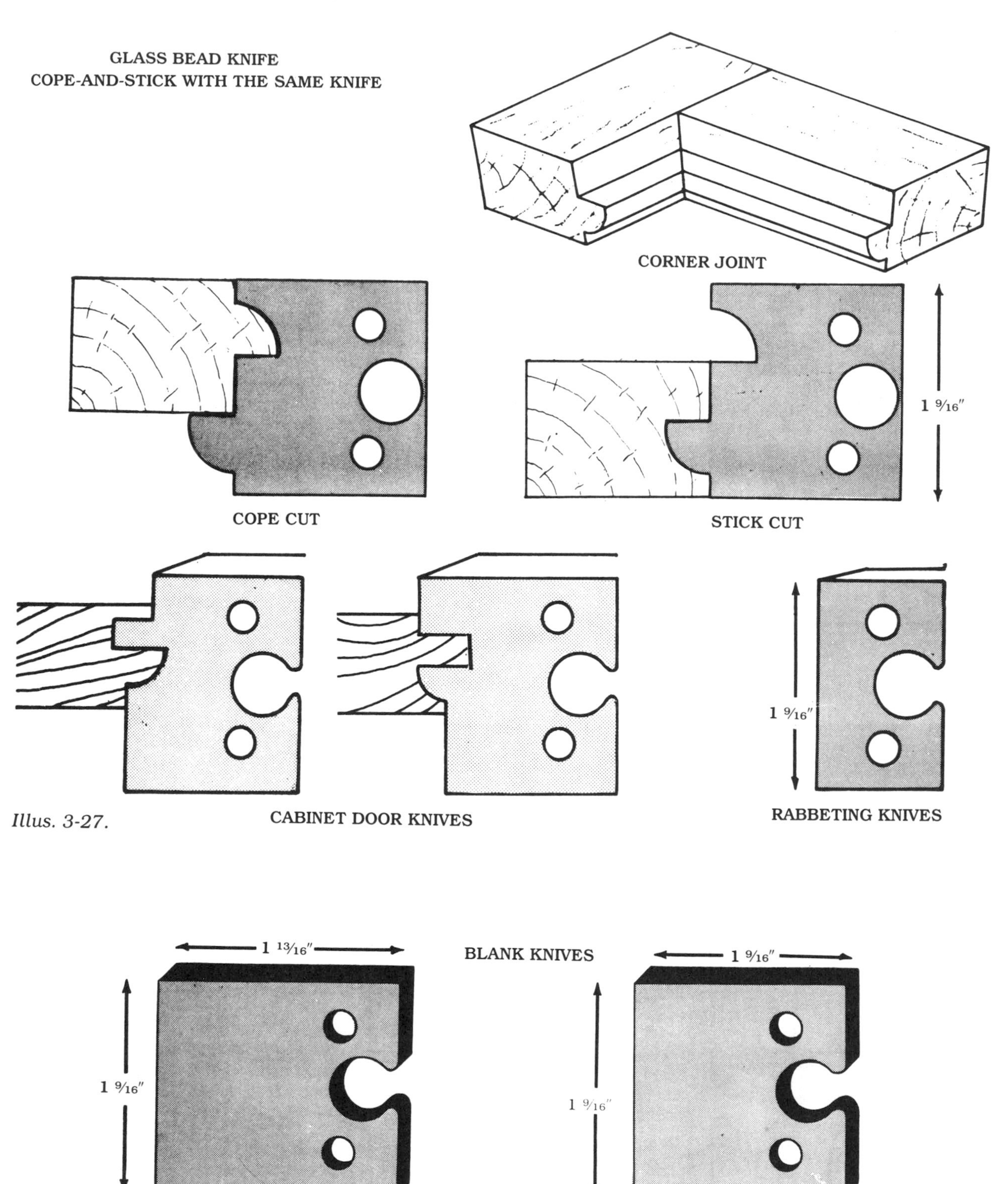

Illus. 3-27.

Illus. 3-28. These knife blanks can be ground to any desired profile. Always have a competent sharpening service grind these cutters for you. (Drawing courtesy of L. A. Weaver Co., Inc.)

Illus. 3-29. The rub collar under the raised-panel cutter rides along the template and controls the depth of cut during this shaping operation. The rub collar or ball bearing can also ride along the edge of the workpiece in certain applications.

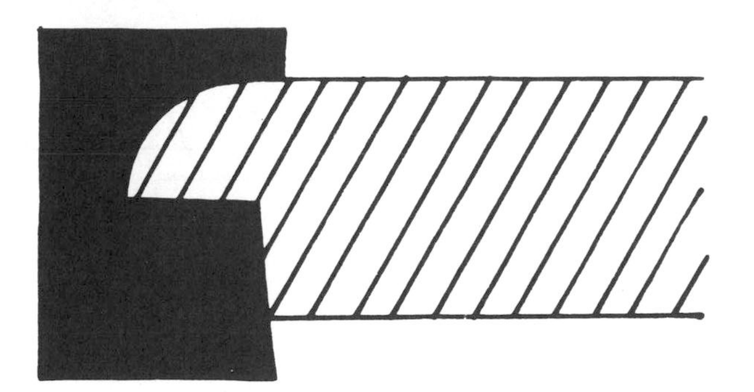

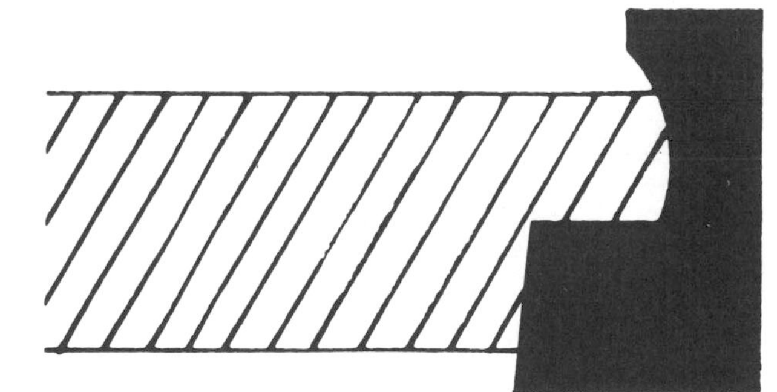

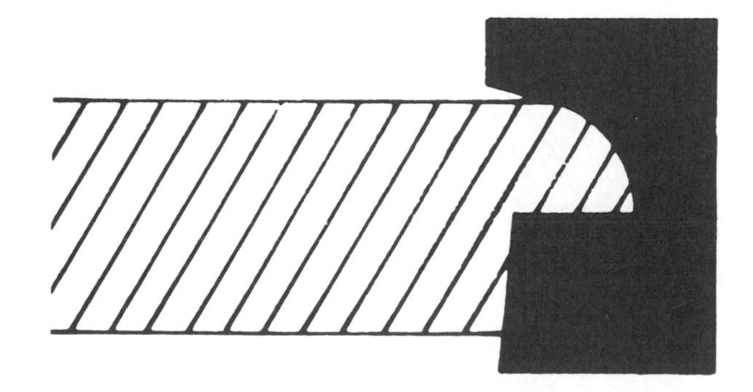

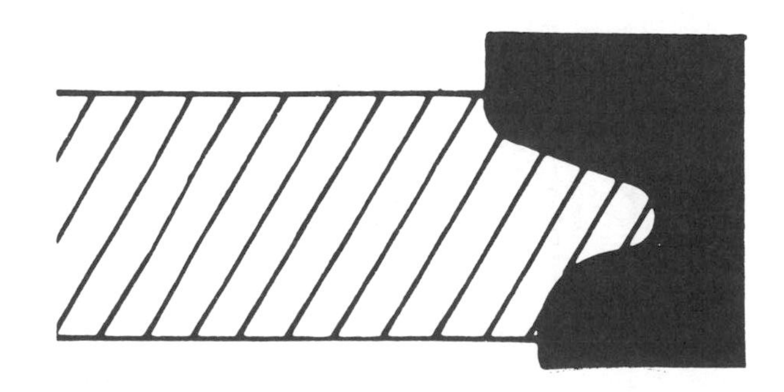

Illus. 3-30. These door-lip and door-edge cutters cut a deep, decorative shape on the edges of cabinet doors. The rabbet formed on the edge of the door compensates for the gap between the rail and stile on the door, and makes them fit much easier. (Drawing courtesy of DML Co., Inc.)

CUTTER PROFILES

Cutters with standard profiles include door edge or lip cutters (Illus. 3-30); glue joint cutters (Illus. 3-31 and 3-32); tongue-and-groove (Illus. 3-33), half-round (Illus. 3-34) and quarter-round (Illus. 3-35) cutters; straight groovers (Illus. 3-36); fluting (Illus. 3-37), roman ogee (Illus. 3-38), and cope-and-stick cutters (which are used to mate door parts) (Illus. 3-39); and various vertical and horizontal raised-panel cutters (Illus. 3-40 and 3-41). These types of cutter have bores ½ inch–1¼ inches in diameter. It is possible to buy all of these profiles mentioned in

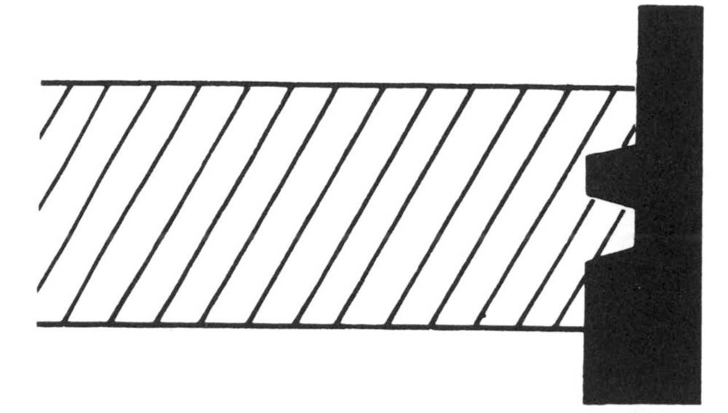

Illus. 3-31. This glue-joint cutter makes a tongue and groove on the edge of boards. These boards can be glued together for greater strength. (Drawing courtesy of DML Co., Inc.)

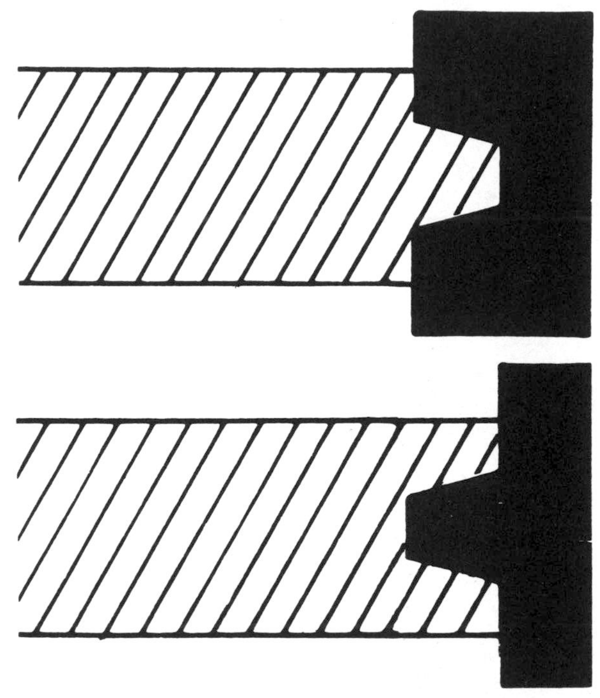

Illus. 3-33. Tongue-and-groove cutters cut a wedge-type joint on the edges of boards so that they can be glued or fitted together. Two separate cutters are used, one to cut the tongue, the other to cut the groove. (Drawings courtesy of DML Co., Inc.)

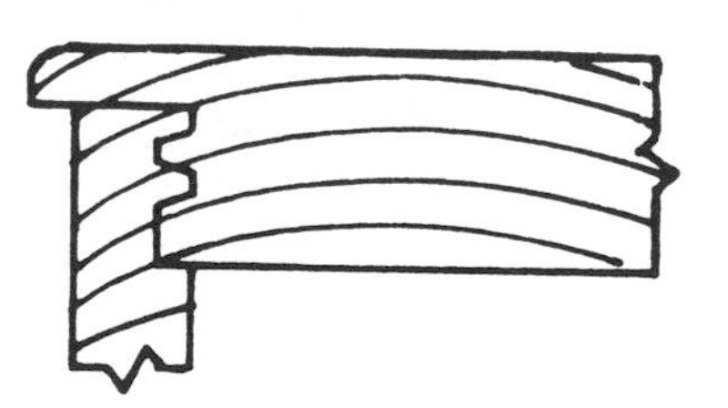

Illus. 3-32. This glue joint cutter can also be used as a drawer-joint cutter. Note the drawing that shows the fit between the drawer front and the drawer side. (Drawing courtesy of the Cascade Tool Co.)

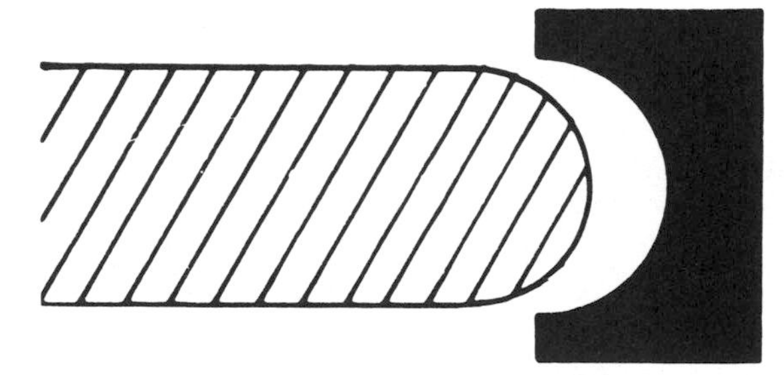

Illus. 3-34. This shaper cutter cuts a half round on the edges of stock. When making smooth contours, remember that the radius of the half round is equal to one-half the stock thickness. (Drawing courtesy of DML Co., Inc.)

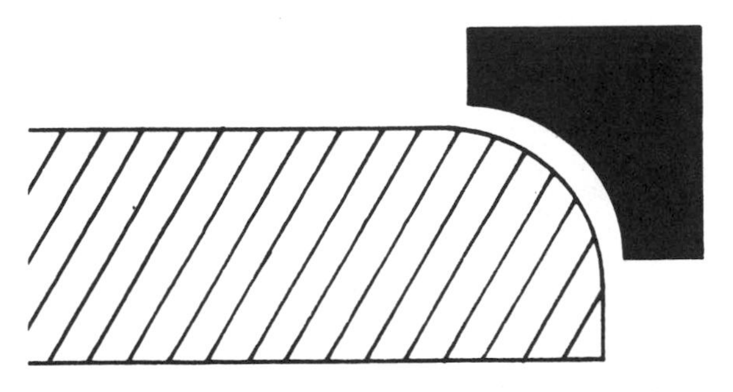

Illus. 3-35. Quarter rounds are radii-shaped cuts made on the edges of boards. In some cases, quarter rounds are used as a moulding between the baseboard and flooring. They are about as wide as they are thick when used in that capacity. (Drawing courtesy of DML Co., Inc.)

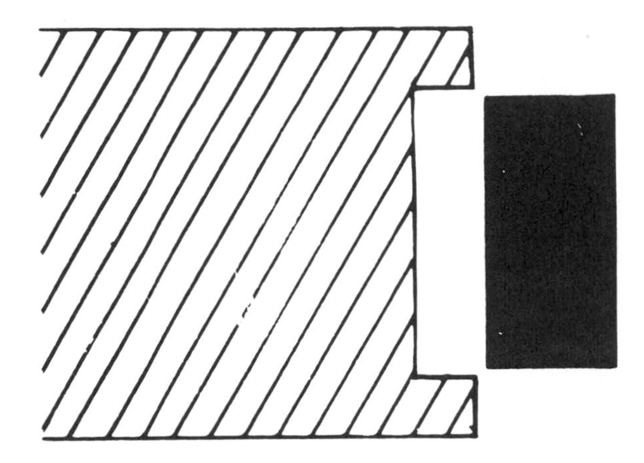

Illus. 3-36. Straight-side cutters or groovers are used to joint the edges of boards or cut a straight square channel along the edges of boards or in the faces of boards. To be effective in cutting rabbets or channels, the top edge and bottom edges of straight-side cutters must have some side clearance. Otherwise, they can only cut the edges of boards. (Drawing courtesy of DML Co., Inc.)

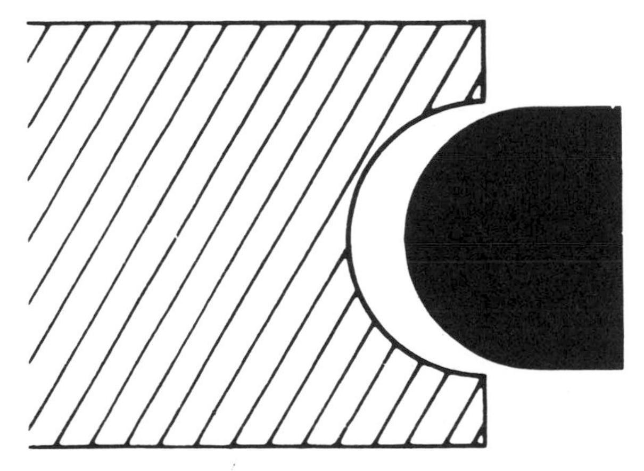

Illus. 3-37. Fluting or radius cutters cut a U-shaped profile in the edges or faces of boards. These profiles are used for decorative purposes in many frames and furniture parts. (Drawing courtesy of DML Co., Inc.)

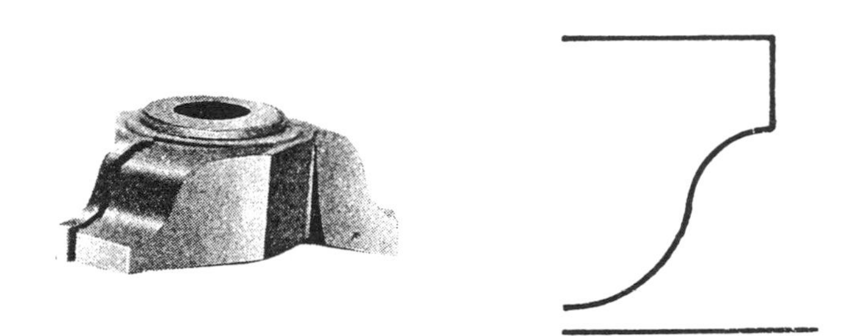

Illus. 3-38. This decorative Roman OG profile makes an interesting edge on furniture parts. (Drawing courtesy of Cascade Tool Co.)

one complete set with the assembled cut arrangement. In fact, there are many sets of solid and assembled cutters that are sold by different manufacturers.

Some cutter sets include the *lock mitre cutter*, which is suitable for many corner joining operations, including the standard edge glue joint. The lock mitre cutter shapes the same profile on two pieces of stock. One piece of stock is shaped with its good face on the table; the other is shaped with its good face against the fence. The two pieces fit together like pieces that have a tongue and groove mitre (Illus. 3-42).

Wedge tongue-and-groove cutters can be best used to align boards. This is because there is a mating fit and a wedging action that occurs during their assembly (Illus. 3-43).

Tongue-and-groove cutters can also be used as glue-joint edge cutters. Straight tongue-and-groove cutters work well, but there is a slight problem when you try to assemble the boards that have been cut. The boards have to be aligned very accurately. This is hard to do because it is hard to make allowances for the application of glue when you cut them. Straight tongue-and-groove cutters are usually used with sets of stacked grooving cutters.

Glue-joint cutters are primarily used for jointing the edge grain of boards, thus providing 30% more glue surface, and for aligning boards

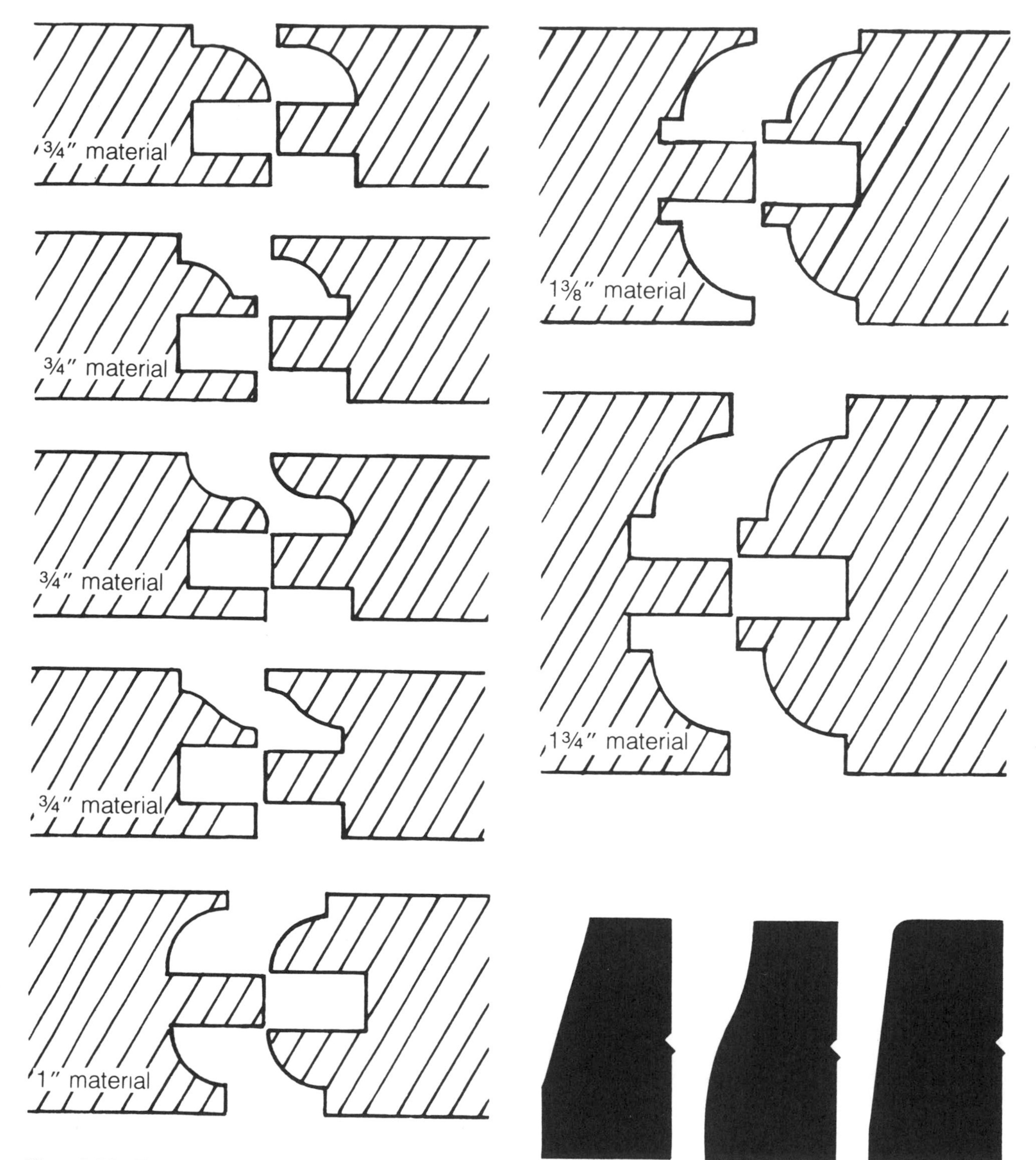

Illus. 3-39. These mating stile- and rail-cutters are sometimes known as cope-and-stick cutters. Note the several profiles that can be used according to the thickness of the workpiece. Some shape a decorative profile on one face of the pieces; others shape a decorative profile on both edges of the pieces. (Drawings courtesy of DML Co., Inc.)

Illus. 3-40. These vertical raised-panel cutters have a smaller orbit, and can be used on lower powered shapers. Note the three different profiles that can be cut. (Drawings courtesy of Delta Industrial Machinery Co. Inc.)

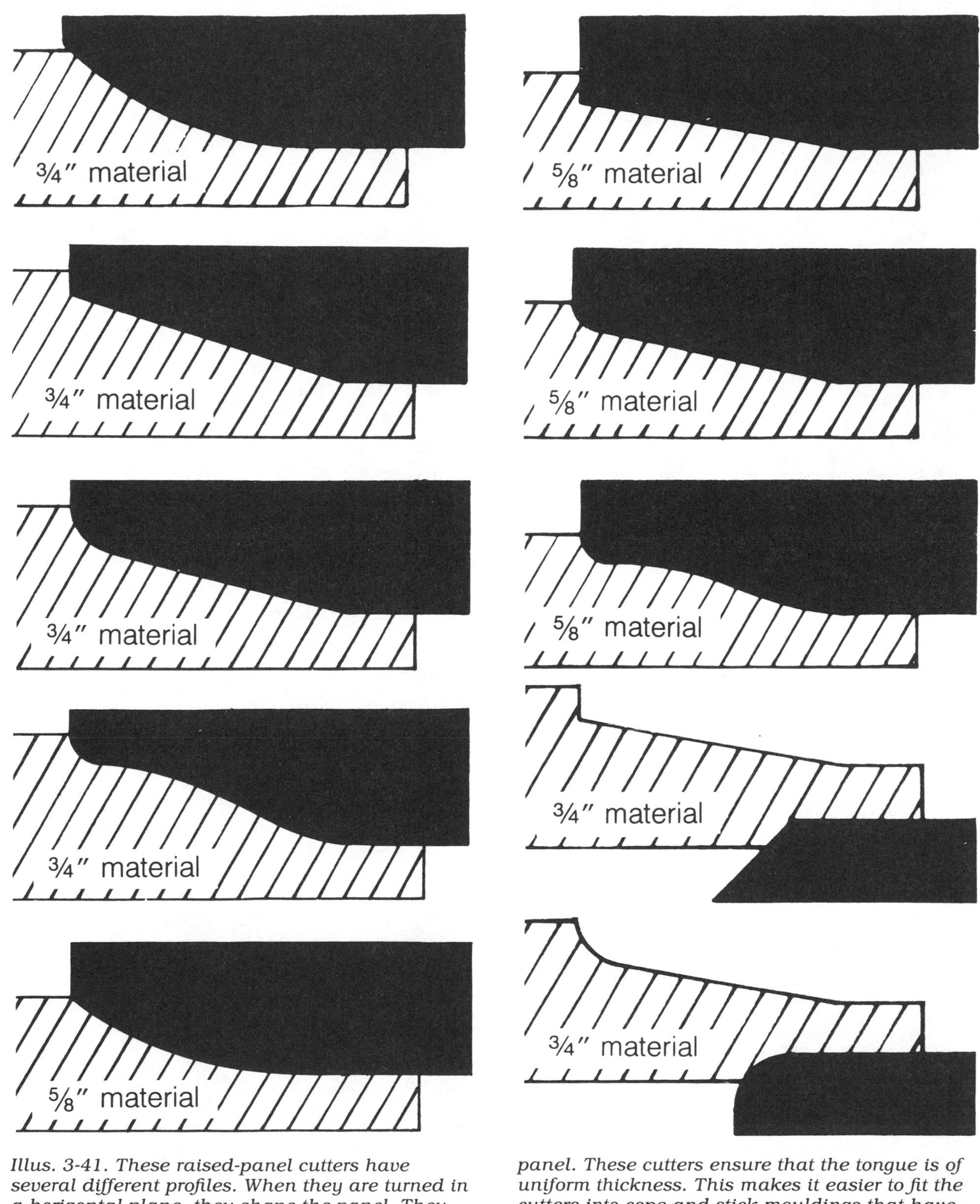

Illus. 3-41. These raised-panel cutters have several different profiles. When they are turned in a horizontal plane, they shape the panel. They can be used with various materials of various thicknesses. In addition, there are special cutters that can be used on the back side of the raised panel. These cutters ensure that the tongue is of uniform thickness. This makes it easier to fit the cutters into cope and stick mouldings that have been already shaped. (Drawings courtesy of DML Co., Inc.)

during the assembly process (Illus. 3-44). To ensure that the boards are perfectly aligned for gluing each time, simply make sure that they are of the same thickness and reverse them either face up or face down before they go through the cutter.

Glue-joint cutters can usually be run either in clockwise or counterclockwise rotations. Glue-joint cutters have thickness-cutting-capacities that range from ½ to 2 inches. The glue-joint cutter utilizes the wedge tongue and groove fit (Illus. 3-45).

Sets of *cope-and-stick cutters* are also available. They also vary according to the thickness of the material being used, as, for example, ¾-inch-thick cabinet doors.

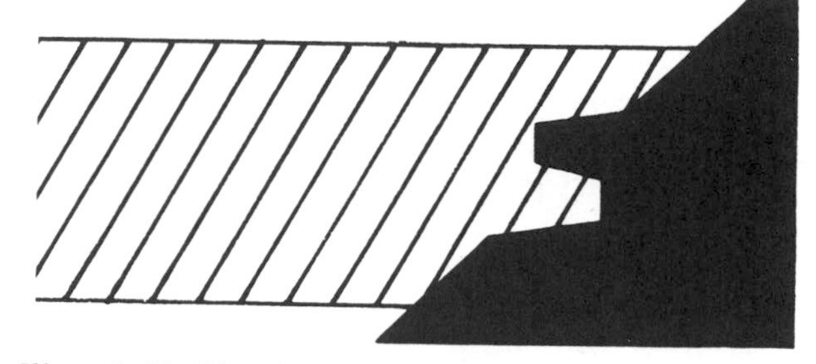

Illus. 3-42. This lock mitre makes a tongue-and-groove joint at a 45-degree angle. This allows the two parts to be fitted together to form a right angle. (Drawing courtesy of DML Co., Inc.)

Illus. 3-43. Tongue-and-groove joints can be used to align boards and fit them together easily. Tongue-and-groove cutters make it easier to align boards into a flat plane.

Illus. 3-44. Glue joints are another form of tongue-and-groove edge joinery. They allow the boards to be aligned into a flat plane, and also provide additional gluing surface.

Illus. 3-45. The joint in the shelf edge was cut with a glue-joint cutter. The face of the shelf edge was shaped after the pieces were glued together. This shaping removed any clamp marks that might have occurred.

Illus. 3-46. These entry-door cope-and-stick cutters cut a decorative profile on both sides of the door. This means that both sides of the door will have a finished look.

Illus. 3-47. Most cope-and-stick cutters come with shim stock that you can use to adjust the space between the cope and stick parts. This allows a tight fit between mating parts regardless of the wood species.

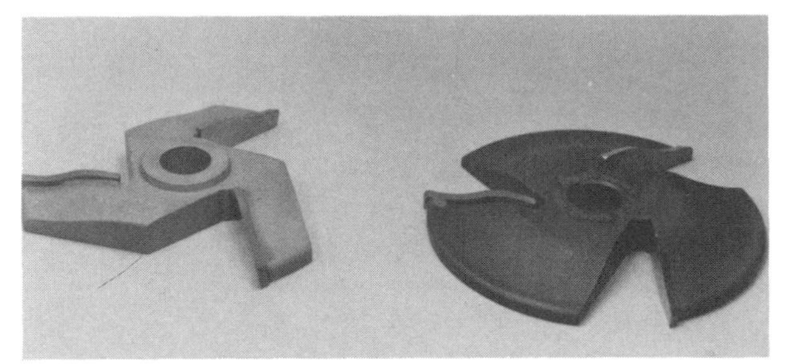

Illus. 3-48. The cutter on the left has an open throat, which allows it to take a big cut or bite as stock is fed into it. The cutter on the right has a closed throat which limits the size of the chip and slows the feed. The cutter on the right is less likely to cause a kickback and is better suited to hand feeding. The cutter on the left is better suited to power feeding, but may be used for hand feeding when light cuts are taken.

Some sets of cope-and-stick pattern cutters have two cutters that are stacked for male and female cuts. Other sets have as many as four or six cutters per male and female cuts. The number of cutters varies according to the manufacturers' designs and the thickness of the door stiles and rails.

Cabinet-door cutter sets are designed to cope a profile on the front of the door only. Entry-door cutter sets cope a finished profile on both sides, resulting in a door that is finished both on its front and back (Illus. 3-46).

There are specialty sets of shaper cutters available that can be stacked to produce cope-and-stick joints for doors, panelling, and many types of frame-and-panel construction. These cutter sets normally come with shim stock that you can use to adjust the fit between tongue and groove as part of the shape (Illus. 3-47).

Never interchange the cutters of different sets. Sets are sharpened together, and must be used together to produce good fitting joints.

You may find when shaping oak with a set of cutters that the pieces fit together nicely, but when you shape willow, the pieces fit too tightly. This may be due to differences in how the wood machines. To get a tight fit in willow, you may have to remove (or add) one shim from the set.

CHIP-LIMITING CUTTERS

Some shaper cutters have chip-limiting shoulders. The chip-limiting shoulder limits or controls the size of the chip that's produced. When

Illus. 3-49. Cutters can be inserted on this cutterhead. This cutterhead has what appears to be two sets of cutters. The second is actually a chip limiter that reduces the size of the chip.

Illus. 3-50. Note the two cutters on each side. The one on the left side of the cutting profile, closest to you, limits the size of the chip. It is actually a few thousandths of an inch smaller than the cutter which is immediately behind it. This reduces the rate of feed and the thrust or kick of the cutter when wood is being shaped.

the chips are smaller, the feed is slower, and the operator has more control over the workpiece.

The two raised-panel cutters shown in Illus. 3-48 show how the chips being cut can affect the job. The cutter on the left has an open throat and will take a big cut or bite. The larger cut increases the chance of kickback and reduces operator control. You can hand-feed this type of cutter if you take light cuts. If you use a power feeder, this cutter will cut freely even if you take deep cuts.

The cutter shown on the right in Illus. 3-48 has a large hub which limits chip size and feed speed, and reduces the chances of kickback. This cutter would be a good choice when you are raising an arch- or crown-top panel by hand-feeding it. The cut produced will be of high quality, and there will be a reduced chance of kickback. If you use this cutter instead of others to make the raised panels for an entire house, you will not become as fatigued.

The cutter used on this cutterhead can be inserted right into it (Illus. 3-49). Closer inspection reveals two "cutters" on each side of this cutterhead (Illus. 3-50). While both appear to be cutters, one is actually a chip limiter. It is a few thousandths of an inch smaller than the cutter. This reduces chip size and thrust, and makes it much easier to hand-feed stock.

This cutterhead is sold as part of a kit (Illus. 3-51). Each cutter has a matching chip limiter (Illus. 3-52).

Illus. 3-53 and 3-54 show other solid cutters that have chip-limiting features.

CUTTERS WITH CUSTOM-GROUND PROFILES

You can make cutters with custom-ground profiles by grinding a solid high-speed-steel blank to the desired profile. Carbide or Tantung™ tips can be added to a steel blank to make a shaper cutter that will stay sharp longer. If you use a

Illus. 3-51. The cutterhead shown in Illus. 3-49 and 3-50 is sold as part of a kit.

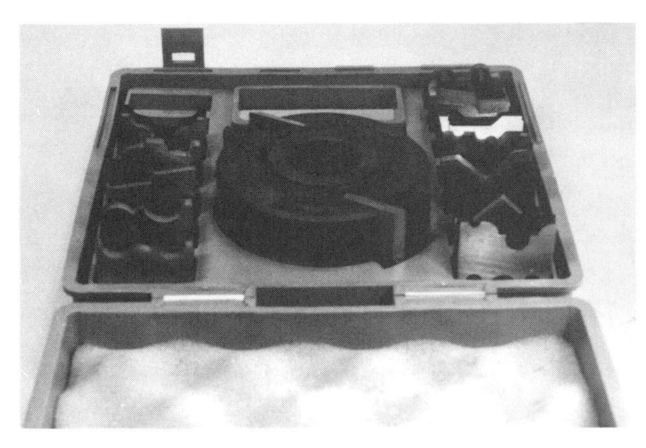

Illus. 3-52. Note the various chip limiters and cutting profiles available that come with this chip-limiting assembled cutterhead. This type of cutterhead is desirable when you are hand-feeding stock.

Illus. 3-53. This solid cutter actually has a chip-limiting profile in it, too. Note the closed throat of the cutters.

Illus. 3-54. This large cutter would have a lot of thrusting power if it didn't have the chip-limiting profiles on it. This makes the cutter much safer for hand feeding. Work carefully when using cutters this large.

shaper lock collar, you can grind solid carbide or high-speed steel knives and fit them to the collar. Always check the manufacturer's instructions to determine the capabilities of your shaper.

You can go to a professional grinder to have a cutter profile custom-made. Provide the grinder with a full-scale sketch of the cutter's profile. This will help him to produce a more exact profile. The profile of your cutter depends on the diameter of the spindle, the number of wings needed for the cutter, and then the direction the cutter will rotate it. You must also determine if the long point of the profile is to be up or down. The cutting circle or cutting diameter of the desired cutter will be determined by the profile needed, as well as by the safety requirements for the shaper as set by the manufacturer. Note: Some profiles require a cutting circle that is too great for the spindle speed of most shapers. Always check with the manufacturer of the cutterhead to determine maximum rpm. Before ordering special cutters from a grinder, check with the shaper manufacturer. Many cutters are available in stock, and will be much less costly.

The material selected for the cutting edge of custom-profile cutters (solid types) depends on the type of wood to be used in the operation, the rpm of the machine being used, and the amount of wood that is to be run through the cutter. If a power-feed unit is going to be used and the rate of feed can be adjusted, the choice of cutting-edge material may depend largely on the cost of the cutter.

Remember that you can easily hone high-speed steel and Tantung for repeated use without changing the profile of the cutter. Though carbide has greater longevity in most cases, it costs most to sharpen.

Accurately ground shaper steel made of M-2 or moly-tungsten is best for average high-production jobs. Always select a piece of steel that is at least ¼ inch wider than the width of the pattern to be cut. Use a knife that is at least one-third as thick as the depth of pattern to be cut. The length of the knife is dependent on the length of the slot in the collar, plus the maximum depth of cut.

Use a piece of steel that's as long as it can possibly be. The heel end of the knife should extend at least to the centerline of the spindle.

Carbide-tipped shaper knives should have at least 3⁄32 inch of carbide. A carbide knife should be at least ⅛ inch wider on each of its sides than the active pattern. A well-designed carbide knife should be at least 5/8 inch deeper than the pattern; this ensures the longevity of the knife and safe operation.

Selecting a Quality Cutter

It is important that you use the correct cutter with your shaper. Check with the manufacturer for exact and safe operating procedures before you buy any cutters. The shaper spindle should have .0002 inch or less runout (eccentricity), and enough horsepower for the size of the profile being removed. Once again, check the manufacturer's specifications.

Remember, if you stack two or more cutters on the spindle, the amount of stock being removed increases. The shaper you are using may not have the power to make a full depth cut when more than one cutter is being used. There is also more spindle deflection, which can affect the quality of the cut.

A power feed is always desirable to use when you are shaping pieces. The cut is more uniform because the piece is being fed at a more uniform speed. The job is also safer, because your hands never get near the cutterhead.

The two-wing cutter is an ideal cutter to use. When selecting a cutter, remember that every tooth must be able to take a big enough bite to dispense the heat caused by the cutting. If the chip is too small, heat builds up in each tooth. This results in stock that has a burned finish and in cutting edges that dull quickly. Burning also makes surface preparation of the wood more difficult. That problem may not be obvious until a stain is applied.

Light, thin cuts due to a cutter with too many wings or because of too slow a feed can cause poor quality work and short cutter life. The cutter cuts the same chip several times, causing premature dulling. A dust collection system will help minimize this problem because the chips are pulled away from the cutter as soon as they are cut.

Carbide tips are much harder than steel tips. They last longer when used on abrasive materials such as plywood, composition boards, and material with glue lines. Carbide, however, will not cut as sharp an edge as high-speed steel. High-speed steel will give the smoothest possible finish when most solid wood is cut.

Illus. 3-55. If the number of knife marks per inch drops below 12, the quality of the finish suffers. The piece shown here was cut at a rate of 6 teeth marks per inch. When this wood is stained, the mill marks will be extremely visible.

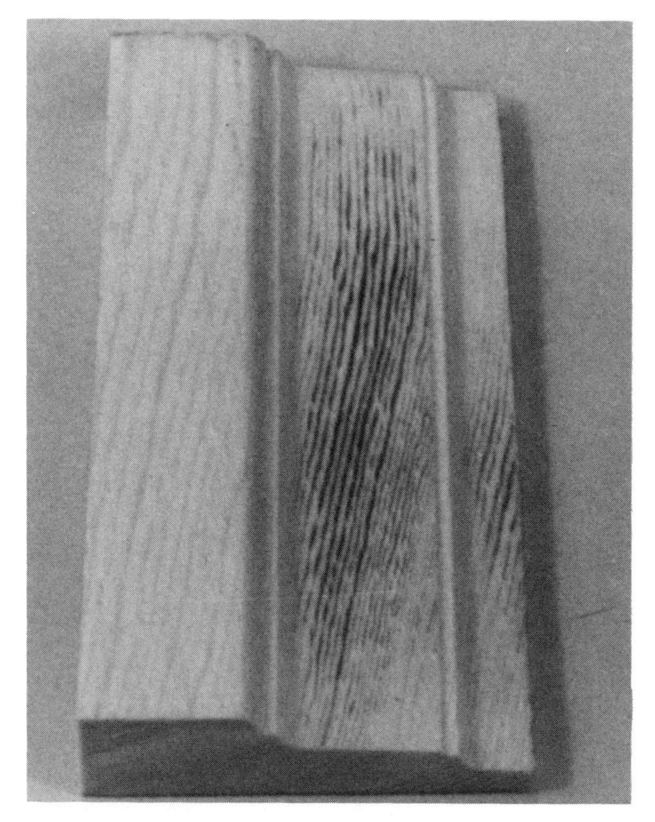

Illus. 3-56. If you cut at a rate that is greater than 30 knife marks per inch, burning can occur. Note the burn mark on this pine. To eliminate burning, make a lighter cut or feed the stock more quickly. In some cases, only a power feeder can feed the wood quickly enough to eliminate burning.

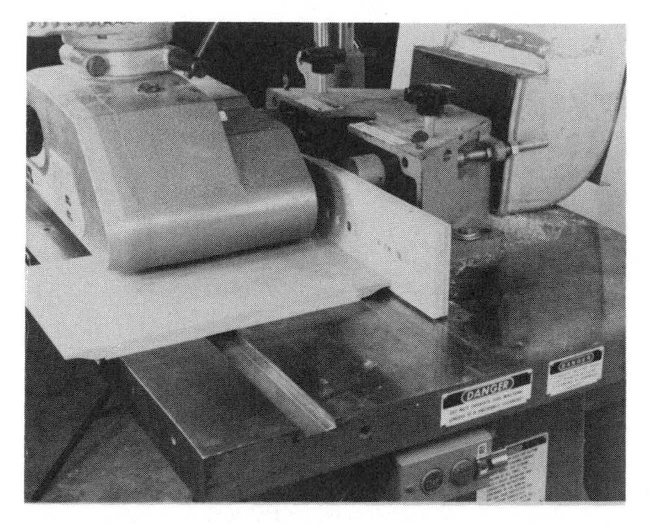

Illus. 3-57. A power feeder can increase the rate at which the workpiece is fed; this eliminates burning when you are shaping large profile in a single pass. This is one of the many advantages of a power feeder.

Two-wing cutters should be used on most hand-feed machines. The desired number of knife marks per inch determines the correct number of wings on the cutter. If the number of knife marks drops below 12, the quality of the finish will suffer (Illus. 3-55). If the number of knife marks exceeds 30, burning and discoloration will occur (Illus. 3-56).

If a machine with a spindle speed of 3,600 rpm is hand-fed, it is fed at rates well under 30 fpm (feet per minute). In such a case, a power-feed unit is very desirable because it will bring the feed rate up to 50 feet per minute. This will result in 12 knife marks per inch (Illus. 3-57). A machine at 7,200 rpm requires a feed rate of over 50 fpm to bring the marks produced down to 12 marks per inch.

Carbide-tipped cutters are usually made with straight-back relief angles because they run freer and cleaner with these types of angles (Illus. 3-58). This type of cutter has the least drag and cuts the most freely. A cutter with a radial-free-back relief angle is used for pattern shapes. The shape will always be maintained if the cutter is sharpened to maintain the original cutting angle. A cutter with a radial relief angle will have lost the minimum amount of diameter as it is sharpened. It is, however, more susceptible to drag or burning. The stock feed is critical for this type of cutter.

Illus. 3-58. The relief angles on these cutters allow them to run freer and make it less likely they will cause burning because there is less drag in the cut.

The best possible way to tell if a cutter is dull is to run a sample board through the shaper (making sure that you are using the proper spindle rpm and feed rate for the wood to watch for burning and drag). Check carefully for any one of the following: missing tips (if carbide cutters are used), chips in either solid high-speed cutters or carbide cutters, any discoloration of tips, and more discoloration of one wing than the other wing(s). (Discoloration in one or more wings suggests heat buildup. Decrease spindle speed, increase the feed rate or do both to decrease heat buildup. Also, never use cutters with chipped or missing cutting wings.)

If the shaper you have has been properly adjusted for rpm and the feed rate is adequate, a sharp cutter will produce a profile that is free of burning and which has 12 cutting marks per inch. The amount of sanding needed will be kept to a minimum.

When cutters are being sharpened, the knives should be ground very slowly and accurately. As a general rule, the amount of material that is ground off should never exceed .003–.004 inch; this ensures that the original bevel is maintained.

Sharpening should be done by a qualified sharpening professional, either available locally or through the cutter manufacturer. Proper grinding for either carbide or high-speed steel should be done using high-quality grinding stones, following the specifications of the cutter manufacturer. All cutting angles must be kept to close tolerances. Some cutters can have a longer life if honed on the flat side of the knife between operations.

Keeping cutters clean also helps to preserve them. Cleaning the cutters removes pitch that has accumulated; this pitch may be acidic. The acid in the pitch can break down the fine cutting edges on the cutter.

Collars

Spacer collars are supplied with the shaper. Spacer collars are used to position the cutters vertically on the spindle (Illus. 3-59). Use as many collars of various heights as necessary to bring the cutter or cutters just below the threads of the spindle.

It is better to use the spacer collars above the cutter, rather than have the cutter at the top of the spindle. The higher the cutter is mounted on the spindle, the more likely it is to deflect and turn eccentrically during operation.

Spacer collars must have the same inside diameter as the spindle being used. Spacer collars vary in thickness from ¼ to 1¼ inches, depending on the manufacturer.

Spacer collars can be custom-made to special height requirements. Custom-made spacer collars are used to position multiple cutters for a desired pattern.

Shim collars are used with spacer collars when 1/1000 of an inch clearance is needed for cutter placement. Shim collars have the same inside diameter as the spindle being used. They can be used with cope and stick cutters to obtain a good mating fit between the cope and stick parts (Illus. 3-47).

The outside diameter of both the shim and the spacer collars should never be greater than the hub of the cutter being used. The hub is approximately ½ inch larger than the diameter of the spindle.

Illus. 3-59. Spacer collars are used to help position shaper cutters vertically on the spindle. They also take up some of the space on the threaded or unthreaded portion of the spindle so that the nut can be securely attached.

Rub or depth collars have ball-bearing centers, and are usually used to control the depth of cut when you are shaping (Illus. 3-60). The rub collar rides along the edge of the workpiece or along the edge of a template or jig.

Rub collars vary in diameter according to the cutters diameter and the desired depth of cut. In some cases, two different rub collars will be used. A larger-diameter depth collar will be used to make the first light cut. A second depth collar of smaller diameter is then mounted for the second cut. This cut will be deeper and will allow the work to be shaped to the desired profile.

In some cases, ball bearings are pressed into metal rings of various diameters. These collars come in different diameters, yet they use the same-diameter ball bearing.

Rub or depth collars are sometimes used in conjunction with fences when the opening between the fences is large. This provides support in the middle of the cut.

Another type of collar is the *dead collar* (Illus. 3-61 and 3-62). Dead collars do not turn with the spindle. They go around the spindle. They limit the depth of cut by riding against the template or workpiece. The dead collar will not cause burning of the template or workpiece. These collars work well when you are raising panels or working with templates.

Illus. 3-63–3-67 show a few of the many profiles that can be cut with the shaper. As you read Chapters 5–8, you will learn how to make some of these cuts.

Illus. 3-60. Rub collars are actually ball bearings that control the depth of cut. They ride along the template or the edge of the workpiece. The washers that you see in front of these bearings go on top and bottom of the ball bearings to allow them to turn freely. Never fit the ball bearings without these washers; they will "seize up" and burn the edge of the workpiece or template.

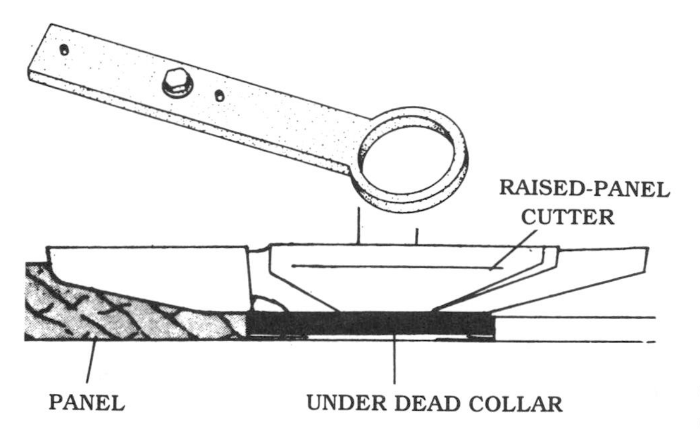

Illus. 3-61. This dead collar controls the depth of cut. It actually goes around the spindle and under the cutterhead. When using the dead collar, be careful when raising or lowering the cutter. It could actually contact the dead collar. The dead collar will never cause burning on the workpiece edge since it does not turn, but remains stationary.

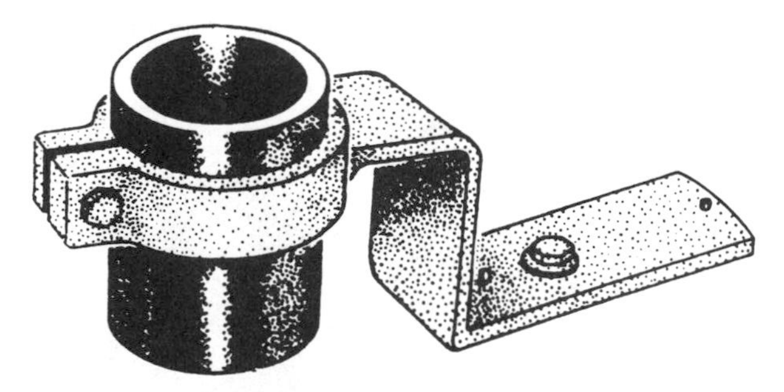

Illus. 3-62. The overhead adjustable dead collar goes over the spindle and actually acts as a barrier between you and the spindle. The template rides against the dead collar; the template is positioned over the workpiece. (Drawing courtesy of L. A. Weaver Co., Inc.)

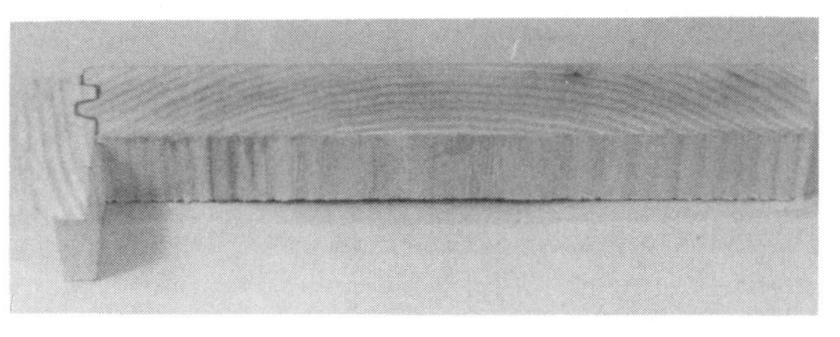

Illus. 3-64. A glue-joint cutter can be used to cut pieces so that they can be joined at a right angle; this makes shelves look thicker than they actually are. This type of cut ensures a perfect fit even if the parts are twisted slightly.

Illus. 3-63. You can bolt this two-piece stair rail to its mating parts before gluing the cap in position. This allows you to finish the railing in the shop and then install it on the job.

Illus. 3-65 (right). After the four pieces were glued up using lock mitre joints, this column was made with 8 cuts on the shaper. Each successive cut contributed to the profiles. The method for making this post is discussed in Chapter 6.

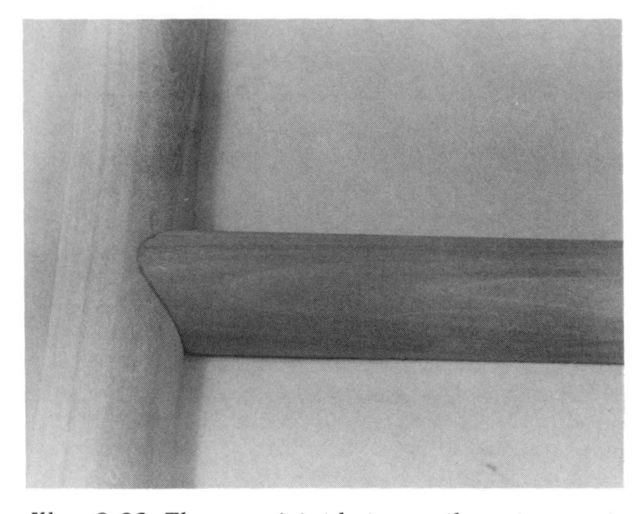

Illus. 3-66. The cope joint between these two parts was actually sanded to eliminate tearout problems. Sometimes shaping or other woodworking techniques require special planning.

Illus. 3-67. Several mouldings actually make up this tabletop. A glass insert is positioned in the middle after the table top is rabbeted. These pieces are a series of six different parts, all glued together and then mitred. This type of work gives the table a unique profile. Techniques for making these cuts are featured in Chapters 6 and 8.

4
SAFETY PROCEDURES

The shaper presents many hazards. There is always the chance that you may accidentally come into contact with the cutters, or that there will be a kickback. Failure to check all hold-downs, clamps, and threaded fasteners can result in a cutter coming loose, or the chance that some object will slide into the cutter.

A safe working environment is the result of safe habits and practices. The goal of this chapter is to help you establish a routine of safe habits and to help identify safe and hazardous conditions. *Therefore, read this chapter in its entirety before performing any shaper operations. Also read it periodically as you continue to use the shaper.*

Causes of Accidents

When an accident occurs, one contributing cause is usually the operator's attitude or his failure to obey all safety regulations. Novice shaper operators may become involved in an accident because they have forgotten to clamp or secure a fence or guard, or because they have failed to identify a dangerous situation.

Experienced shaper operators usually become involved in an accident because they disregard safety practices. As the experienced operator becomes more confident, there is a greater likelihood that he will take chances or disregard a safety practice. When he gets away with the unsafe act, it is likely that it will be repeated. As a woodworker, it is good practice to adhere to safety practices and habits at any cost.

Most accidents have other contributing factors, among which are the following:

1. *Working while tired or taking medication.* Whenever you are tired, stop or take a break. Accidents are most likely to happen when you are tired. Medication and alcohol can affect your perception and reaction time.
2. *Rushing the job.* Trying to finish a job in a hurry leads to errors and accidents. The stress of rushing the job also leads to early fatigue. When you find yourself rushing, ask yourself how fast the job will be completed if you get hurt or damage some of the work.
3. *Inattention to the job.* Daydreaming or thinking about another job while operating the shaper can lead to an accident. Repetitive cuts lend themselves to daydreaming. Be doubly careful during production shaping operations.

   Setting up the shaper also requires your full attention. There are many adjustments to be made when setting up the shaper. Daydreaming can lead to an incorrect setup and possible accident or injury.

   Always double-check a setup before you begin shaping. Develop setup routines that become habit. This reduces the chance of error.
4. *Distractions.* People talking in the workshop, unfamiliar noises, and doors opening and closing are all distractions in the shop. Shut off the shaper before you converse with someone, look up, or investigate an unfamiliar noise. When you wish to approach coworkers operating a shaper, make sure that they see you before you address them. Startling them can cause an accident.
5. *General housekeeping.* A dirty or cluttered work area provides tripping hazards and excess dust that can be a breathing hazard. Use a dust-collection system when using the shaper (Illus. 4-1). Keep the shop neat and clean. It is more pleasant and safer to work in a clean area.

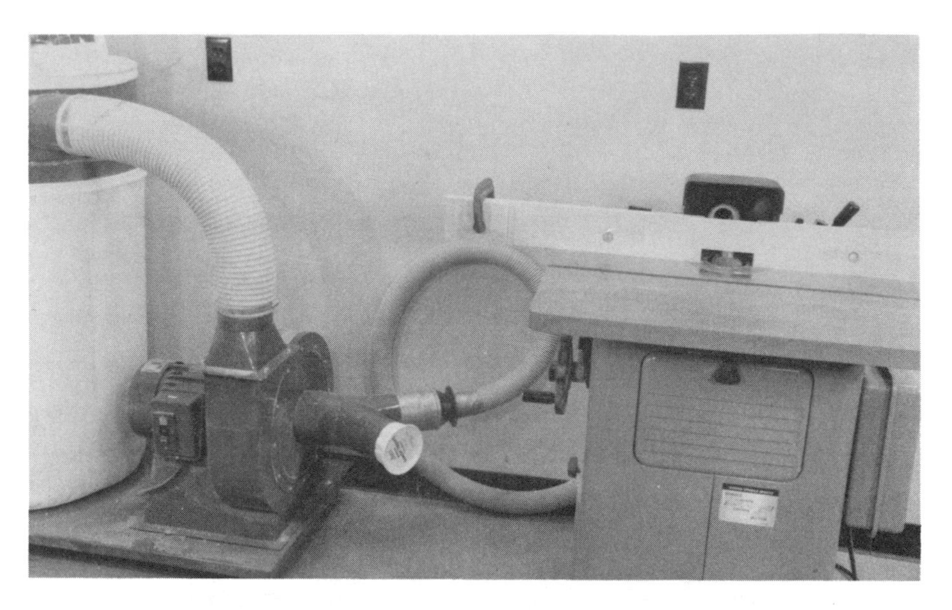

Illus. 4-1. A dust-collection system makes it safer to use a shaper, and also improves the quality of the cut. The chip is only cut once, and then it is drawn away.

Illus. 4-2. This Shop-helper wheel pulls the workpiece down against the table and against the fence during shaping. The wheels turn in only one direction. This minimizes the chance of kickback and improves the quality of the cut.

Illus. 4-3. These spring hold-downs reduce vibration during cutting and also hold the stock firmly on the table. Some of these hold-downs can also be used to push the stock in against the fence. Check the manufacturer's directions to learn how to use the hold-downs correctly.

Illus. 4-4. The device shown here acts as a hold-down and as a barrier between the operator and the cutter during shaping. If you tighten the nut at the top of this device, you can increase or decrease the tension on the wooden shoe. This also helps hold the stock in position, and gives you greater control of the workpiece during shaping.

Illus. 4-5. Once the tension is adjusted, the hold-down keeps the stock on the table and reduces vibration.

Kickbacks

A kickback occurs when a piece of stock is forced towards the operator at great speed. In most cases, the shaper cutter hits a knot and kicks the work back. In other cases, the workpiece gets pinched between the cutter and the guard. This is likely to cause a severe kickback.

Stock must be under control at all times. Stock can kick back at great velocity. This is a serious hazard. Another hazard of the kickback is the fact that the operator's hand may be pulled into the cutterhead as the stock kicks back.

You can minimize the chances of kickback by observing the following precautions:

1. Shape only true, smooth stock that will not twist into the cutter.
2. Use stock-controlling devices such as hold-downs or power feeders to keep stock snug against the fence and table (Illus. 4-2–4-8).
3. Use only sharp cutters. Dull or pitch-loaded cutters lend themselves to kickbacks.
4. Whenever possible, cut the face or edge of the work closest to the table. When the cutter is under the workpiece, the workpiece will not get pinched between the cutter and table.
5. Control all cuts with a mitre gauge, fence, rub collar, template, or a combination of these devices (Illus. 4-9–4-12). Never attempt to shape stock without a control device. Failure to control stock could allow it to be grabbed by the cutters and kicked back towards you.
6. Always feed the piece being shaped completely past the cutterhead. Never release the work while it is still touching the cutterhead.

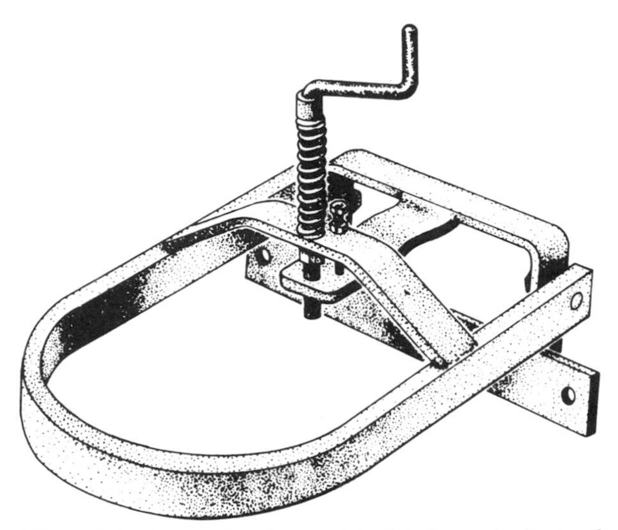

Illus. 4-6. This raised-panel hold-down bolts to the shaper and provides tension or holding pressure on the top of the workpiece as the panel is raised. You can adjust this device by turning the crank at the top. Set the tension to correspond with the amount of cut being taken. (Drawing courtesy of L. A. Weaver Co., Inc.)

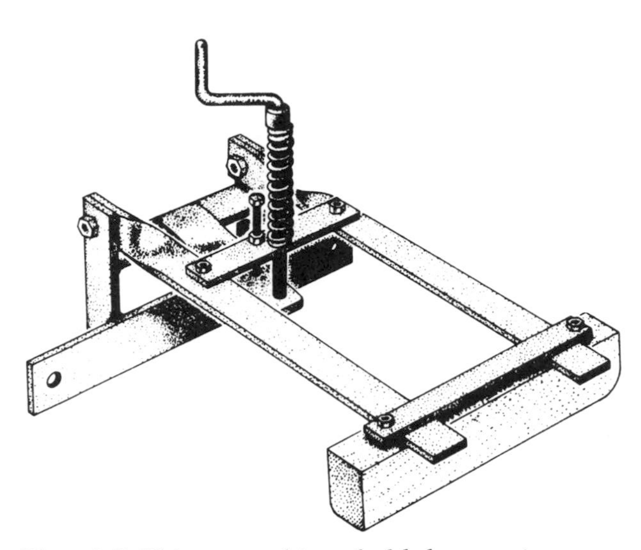

Illus. 4-7. This general-type hold-down acts as a tension-generating hold-down for the workpiece in general shaping operations. You can increase or release the tension by cranking the handle on the top of the device. (Drawing courtesy of L. A. Weaver Co., Inc.)

A kickback may result. Use a push stick to control thin or narrow strips (Illus. 4-13–4-16).

7. Stand to the side of the work while shaping. If the work were to kick back, you would be clear of it.
8. Keep fences close to the cutter. The smaller the opening between the fences and cutter, the less chance of a kickback (Illus. 4-17 and 4-18).
9. Install a table insert that helps support the work close to the cutter.

Cutterhead Safety Procedures

When working with the shaper, you will be required to change cutters frequently. Some cutters will be solid two- or three-wing cutters. Others will be two-wing cutters that are assembled into a cutterhead.

When you install solid three-wing cutters, make sure that they are sharp. Inspect them carefully. The cutting surfaces should come to a point. There should be no radius here.

To determine if the cutter is sharp, drag your fingernail across the cutting edge. The tool should raise a chip on your fingernail. If it doesn't, it is probably too dull to cut effectively.

Always make sure that the cutter is mounted properly. The flat side of a three-wing cutter should point towards the side of the table from which you plan to feed stock. On reversible spindle shapers, this is not a problem. If the spindle turns in only one direction, the cutter has to be positioned with this in mind.

Assembled cutters are designed to be used in a specific cutterhead. Select the cutters according to the cutterhead you plan to use. To work safely, make sure that the cutters are equal in weight and dimensions (Illus. 4-19). This allows for a balanced operation (Illus. 4-20) and uniform clamping.

Some control-cut cutterheads have matching cutters and chip breakers or limiters. In this case, install all these parts in the cutterhead. The chip breakers limit feed speed and reduce the chance of a kickback (Illus. 4-21). They also reduce the depth of cut. This reduces the chance of kickback and makes hand feeding safer.

Check the cutters before installing them. Make sure that they all have the same profile. Mount the cutters according to the manufacturer's directions.

Illus. 4-8. This joining setup uses a power feeder for a hold-down. Stock is controlled by two fences, one on each side of the workpiece. The feed is controlled by the three wheels or rollers on the power feeder. The power feeder guides the work and acts as a barrier between the operator and the cutter. (Photo courtesy of L. A. Weaver Co., Inc.)

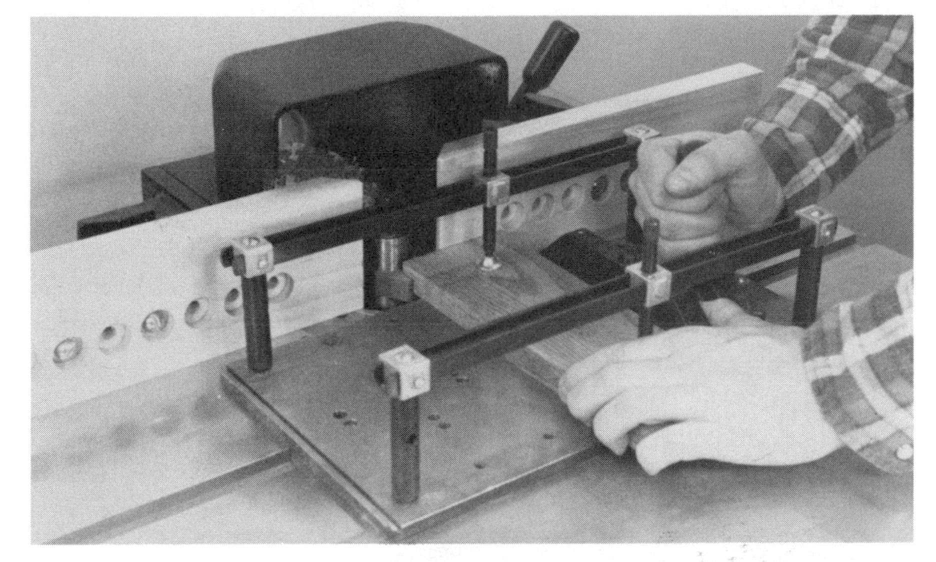

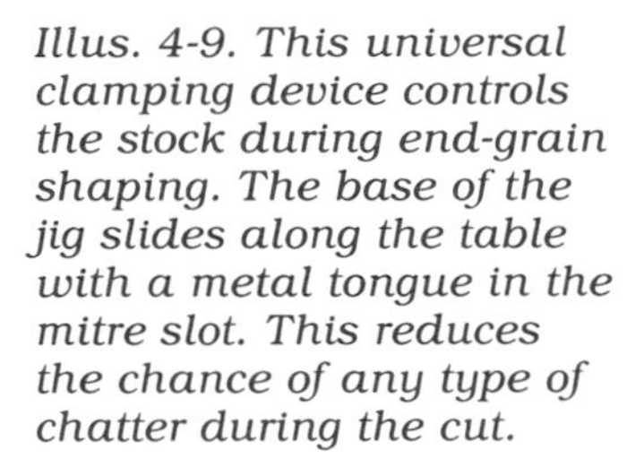

Illus. 4-9. This universal clamping device controls the stock during end-grain shaping. The base of the jig slides along the table with a metal tongue in the mitre slot. This reduces the chance of any type of chatter during the cut.

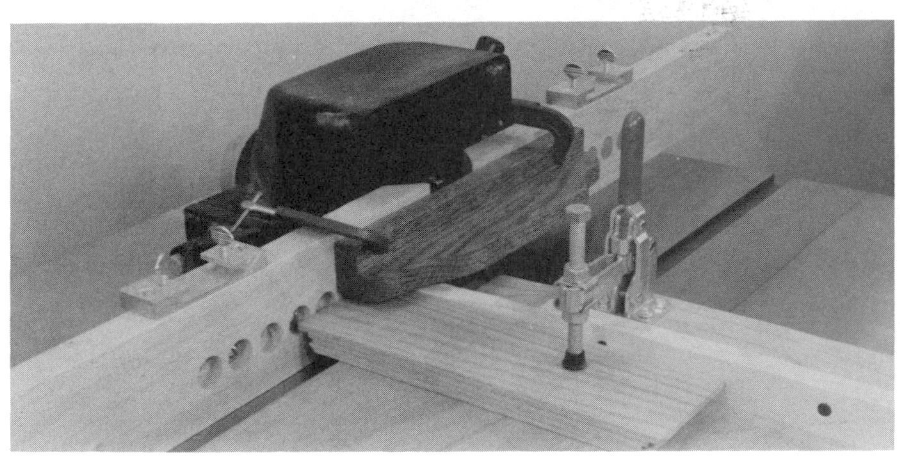

Illus. 4-10. This shop-made holding device actually works off the mitre slot. It uses an industrial clamp to secure the piece to the holding jig. Note the barrier guard above the shaper cutter.

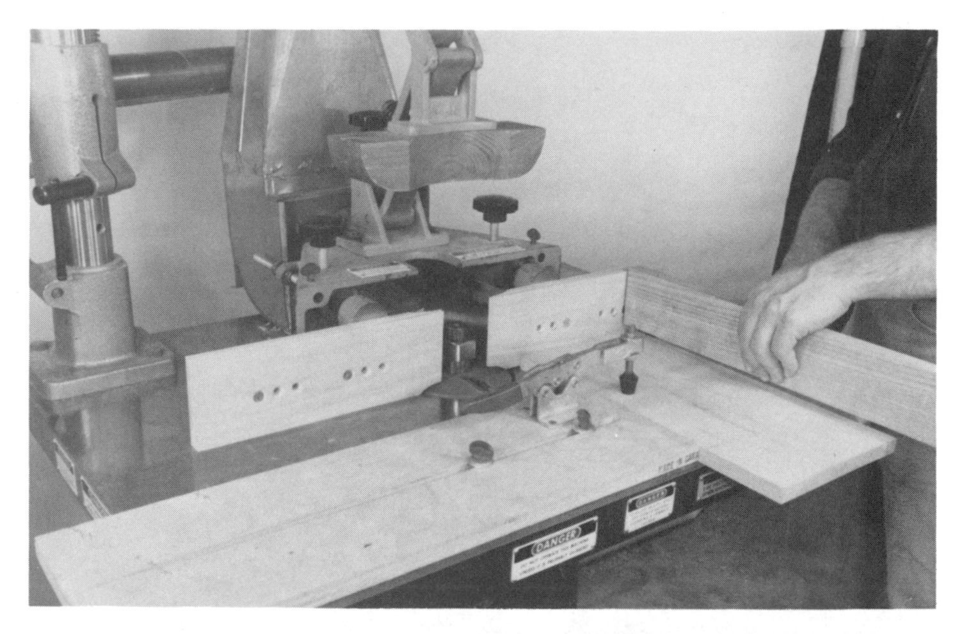

Illus. 4-11. This specialty setup was designed to secure the workpiece during raised-panel cutting. The panel is held securely in the mitre gauge, and the industrial clamp, which is mounted on the tongue of the mitre gauge, holds the workpiece down against the table.

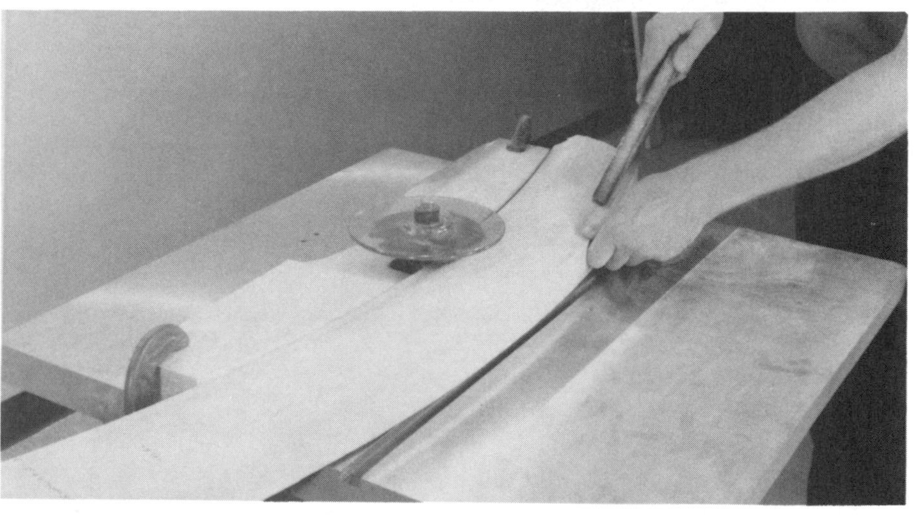

Illus. 4-12. This curved fence acts as a controlling device for the workpiece. There is also a rub collar on the underside of the panel raiser used to shape this railing. This collar acts as a depth guide. Note the safety barrier above the workpiece that minimizes the chance of contact with the cutter.

Illus. 4-13. A push stick can be used to guide or control work being shaped. The push stick keeps your hands a safer distance from the cutter. Always use push sticks when working with thin or narrow strips.

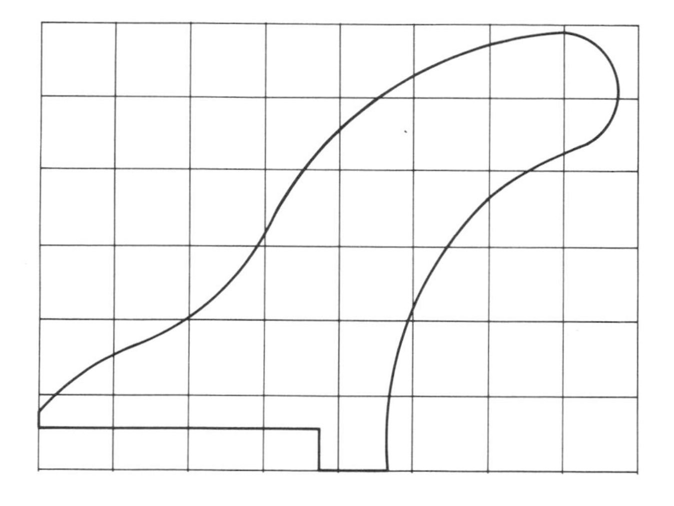

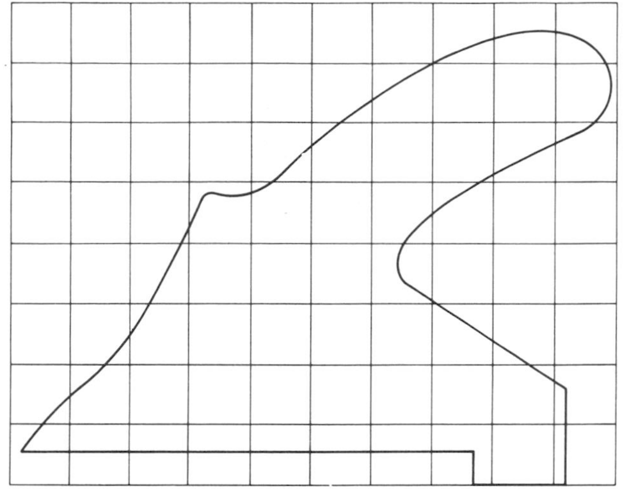

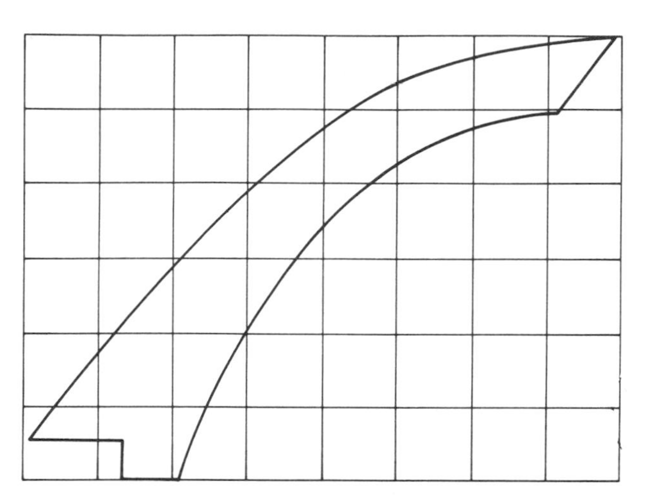

Illus. 4-14–4-16. Use these drawings as guides for making push sticks. Cut the push sticks from 3/4-inch plywood and radius the edges of the handle. When the push sticks become nicked, throw them away and make new ones. One square in the drawings is equal to one inch.

Illus. 4-17. Note that the opening between the fences is limited to accommodate just the radius of the cutter. Always keep your fences close to the cutter, even if it means cutting away a portion of the workpiece to accommodate other safety devices. The smaller the opening between the fences, the less likely the chance of a kickback.

Illus. 4-18. Note the large snipe at the end of this workpiece. It was caused when the piece fell into a large opening in the fences. In many cases, this would cause a kickback rather than snipe the workpiece. Always limit the opening between the fences.

Illus. 4-19. When inserting a cutter into a cutterhead, insert it into more than half the length of the groove it is supposed to be held in. Always limit the amount of cutter being exposed according to manufacturer's directions. Check these cutters periodically to be sure that they are securely fastened in the cutterhead.

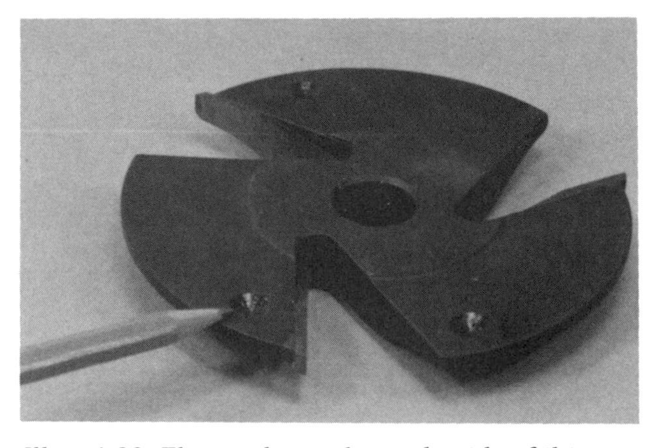

Illus. 4-20. The marks on the underside of this cutter suggest that it has been balanced for better operation. For safety purposes, cutters should be balanced. This reduces vibration and chattering during cutting.

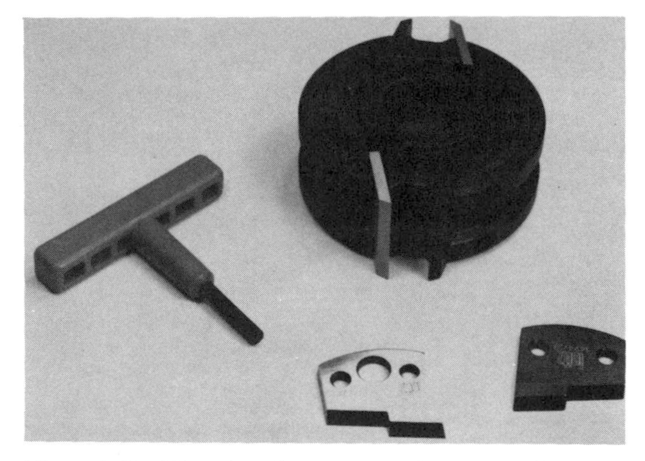

Illus. 4-21. The chip limiters on this cutter limit the feed speed and reduce the chance of kickback. Chip breakers are discussed thoroughly in Chapter 3.

Some solid cutters have small gullets. Gullets are the space between the cutters and the cutterhead. A small gullet limits the depth of cut. This means that a kickback is less likely and that hand feeding will be much safer. Chapter 3 discusses this aspect in detail.

Regardless of the cutterhead you are using, make sure that you lock the cutters securely into position. It is good practice to check them periodically as you work. Any vibration in the shaper should be an indication that the cutterhead is not balanced correctly. Stop the machine at the first sign of vibration. Inspect the cutterhead and the spindle for possible causes of vibration.

It may be necessary to check the spindle with

a dial indicator. This will reveal any eccentricity or runout at the spindle.

Replace the spindle if there is any indication of spindle runout. Check the collars and spindle nuts for burrs. These problems could cause the spindle to bend slightly when the spindle nuts are tightened.

Make sure that the cutters you plan to use have been designed to cut the material you plan to shape. When tool-steel shaper cutters are used on fibre-core plywood, they do not stay sharp very long. This could damage the cutters and result in a greater chance of kickback. When in doubt, check with the cutter manufacturer to be certain that the cutter will handle the job in question.

Proper Setup Procedures

Setting up a shaper is a complex series of settings and adjustments. For the sake of safety, it is wise to break the job down into identifiable parts, such as: attaching the shaper head to the spindle, adjusting the cutter height, adjusting the fences, positioning the guards, etc.

Concentrate when setting up the shaper. Develop a list of shaper settings and check off each setting as it is completed. Use this list as a final check.

Go through all settings a second time. This approach may seem repetitious, but it assures the operator that no mistake has been made. This is very important when the person setting up the shaper cannot complete the setup at one time. He may not be able to complete the setup because he is being interrupted in his workshop while doing it, or because he is a hobbyist who can only set up the shaper after work, and needs two or three nights to do the job. Without a checklist, an error could easily be made in the setup.

Be certain that everything you adjust is locked securely. If anything moves during the shaping operation, stock could be wasted or an accident could occur. Any time spent checking a setup is time well invested.

General Working Environment

The general working environment can also be a factor in the safe operation of the shaper. The shaper should be set at a comfortable height. Most operators prefer a height of about 34–38 inches. Be sure the shaper has been levelled properly and that it does not rock (Illus. 4-22). Whenever possible, it is best to secure the shaper to the floor (Illus. 4-23 and 4-24).

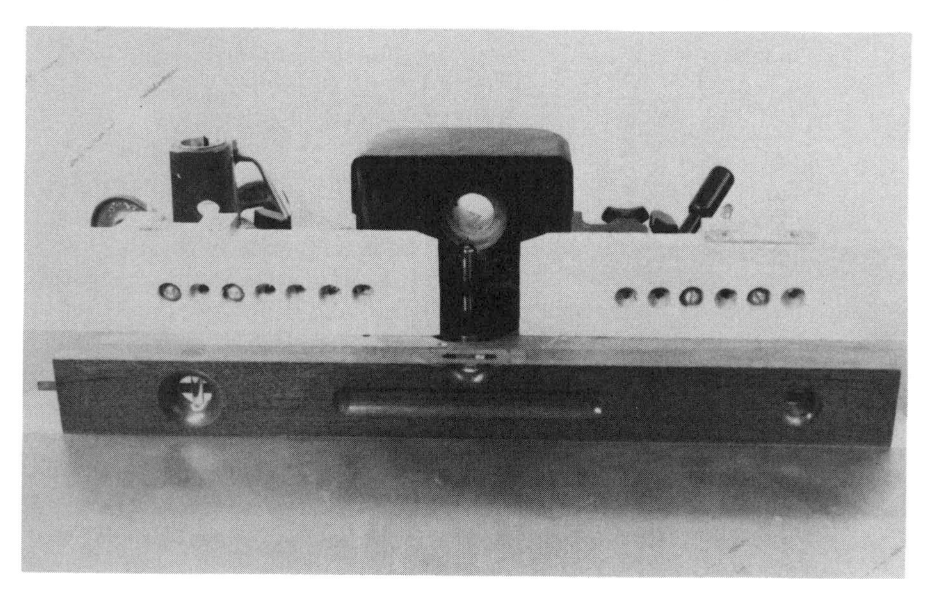

Illus. 4-22. Before shaping, make sure that the shaper is level and securely balanced on the floor. The shaper base should not rock at all. Rocking generates excess vibration.

Illus. 4-23. This shaper has been bolted to the floor. It is a 10-horsepower machine and requires absolute rigidity for the large profiles that it shapes. The pedal at the base is an emergency brake which stops the motor and the spindle instantly.

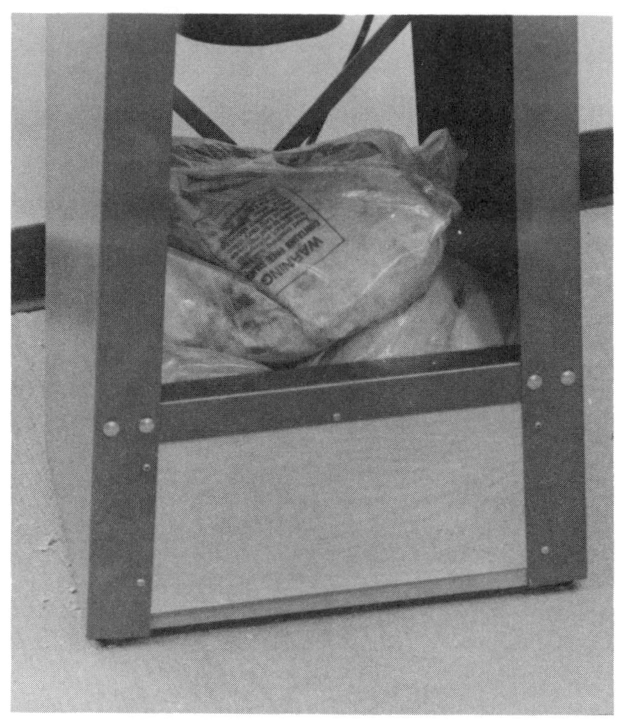

Illus. 4-24. The base of this shaper has been filled with bags of sand to anchor it to the floor. This ensures that it does not move or vibrate during shaping. Some shapers are top-heavy. This means that during a heavy cut they could actually tip over if extra weight were not added to their bases.

Make sure that the appropriate outlet is conveniently located. It should have the correct voltage and amperage. The circuit should be wired and grounded correctly. The outlet should be located below the shaper. This keeps the cord from interfering with the work. It also reduces the chance that you will trip over the cord.

Shaper Operating Rules

Following are rules you should adhere to when using a shaper:

1. *Wear protective clothing.* Always wear protective glasses when you operate the shaper (Illus. 4-25). If the work is noisy, wear earplugs or muffs to preserve your hearing and minimize fatigue.

You can wear gloves when handling rough lumber, but never wear them (or loose clothing) when operating the shaper. Your hand could easily be pulled into the cutterhead if the glove were caught on the cutterhead.

2. *Use the appropriate guards and hold-downs.* Guards act as a barrier and limit contact with the cutterhead (Illus. 4-26 and 4-27). Without the guard, the slightest slip could cause a serious injury. In all cases, a guard should be used to prevent the operator from coming into contact with the cutterhead (Illus. 4-28).

Adequate lighting also makes operating the shaper much safer. Shadows and dim lighting increase operator fatigue and measurement errors.

The area surrounding the shaper should be large enough to handle big pieces of stock. Traffic should be routed away from the kickback zone. The floor surrounding the shaper should be kept free of dust and debris. They can be a slipping or tripping hazard.

Illus. 4-25. Always wear protective glasses and a face shield if possible during shaping. This will minimize the chance of injury during shaping. If you are in a noisy environment, wear ear-plugs or earmuffs to preserve your hearing and minimize fatigue.

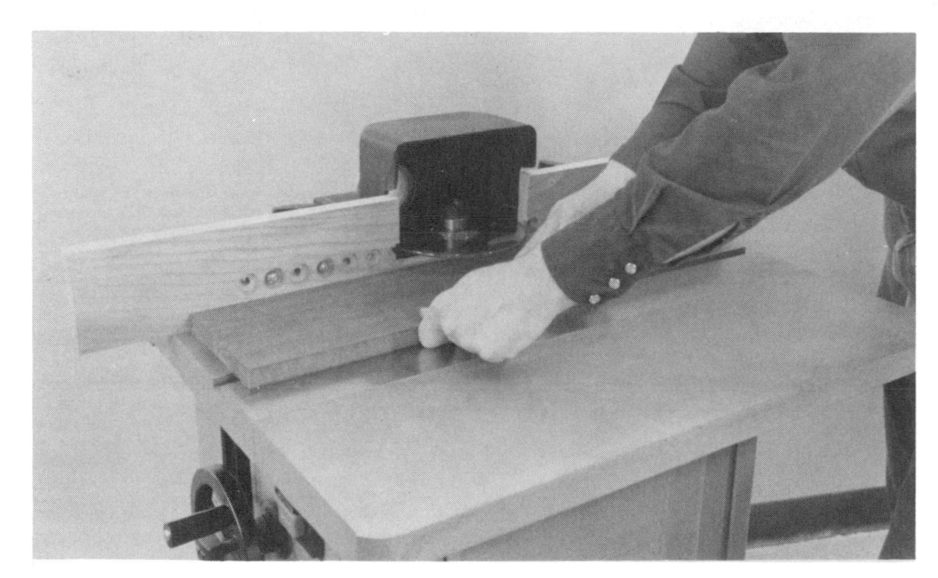

Illus. 4-26. This plastic guard acts as a barrier between the cutter and the operator. It has a ball-bearing center and is actually mounted to the spindle. It is positioned just above the workpiece.

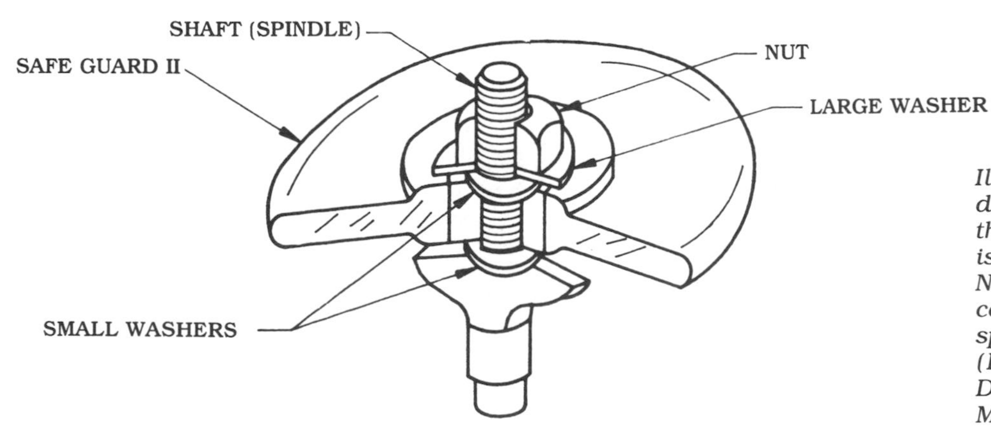

Illus. 4-27. This line drawing shows you how the Safe Guard II barrier is mounted on the spindle. Note the ball-bearing center that allows it to spin freely on the spindle. (Drawing courtesy of Delta International Machinery Corp.)

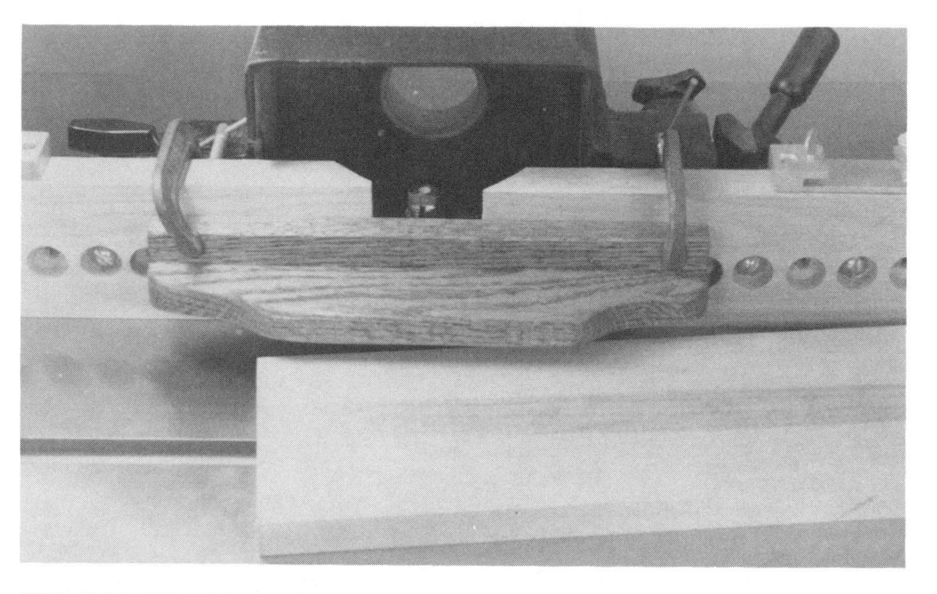

Illus. 4-28. This shop-made barrier will eliminate contact between the operator and the cutter. A barrier like this can be made in the shop in a minimum amount of time. To work safely, one should have a series of these barrier guards.

Illus. 4-29. This power feeder, working in conjunction with the hold-down and barrier guard, makes the operation very safe. The operator's hands are never near the cutterhead. The work is always fed at the ideal speed, and vibration is controlled with the heavy weight of the power feeder. (Photo courtesy of L. A. Weaver Co., Inc.)

Illus. 4-30. If you keep the unused portion of the shaper cutter beneath the table, the workpiece actually acts as a barrier between you and the cutterhead. Note also the barrier guard above the workpiece, which limits contact with the spindle.

Illus. 4-31. Check the cutters periodically to be sure that they are sharp. Also be sure that the spindle is mounted securely and according to the owner's manual.

There are numerous guards and guarding configurations. Use the one most appropriate for the operation you are performing.

Hold-downs also make shaping safer. They offer positive control over the workpiece and minimize the chance of kickback. Hold-downs vary in size and shape, but they all help control stock.

Power feeders act as a hold-down (Illus. 4-29). They are electrically powered rollers that pull the stock into the cutter. The rollers control the stock and, in some cases, act as a barrier to the cutter. Power feeders also keep the operator's hands clear of the shaper cutters.

It should be noted, however, that while a power feeder makes shaping safer, it presents a danger of its own. If your fingers are under the workpiece as it is grabbed by the rollers, it could catch and pinch them. Handle the stock carefully when using a power feeder.

3. *Minimize the amount of cutter exposed.* Keep any unused portion of the cutter beneath the table when shaping. By keeping the unused portion of the cutter beneath the table, you further reduce the chance of contact with the cutters (Illus. 4-30). In addition, by shaping the underside of the work, the stock is not pinched between the cutter and the table. The workpiece acts as a barrier between the cutter and you. This minimizes your chance of contact with the cutterhead.

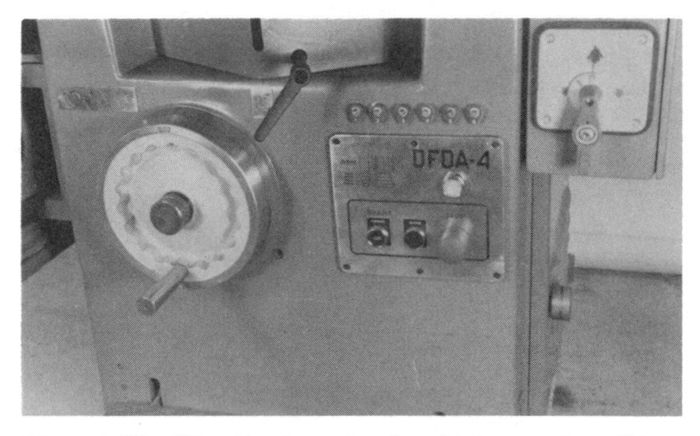

Illus. 4-32. All adjustments should be made with the power disconnected. This shaper features a locking device in the center of the off/on switch, so it can be locked off while adjustments are being made. This locking device also minimizes the chance that an untrained person will use the shaper.

4. *Keep the cutterhead sharp.* A sharp cutterhead makes the shaper much safer to use. A dull cutterhead increases the chance of kickback. A dull cutterhead also requires increased cutting force. This excess force can throw the operator off balance and lead to an accident.

Check the cutters periodically to be sure that they are sharp (Illus. 4-31). Do this with the power disconnected. Have the cutters sharpened at the first signs of dullness. Chapter 3

Illus. 4-33. Always inspect the stock before performing any shaping operation. Stock with loose knots, twists, or cupping should not be shaped because there is a chance it will be kicked back. Note the dead collar mounted on the table which acts as a depth-control device. This does not turn, but rather wraps around the spindle and provides a bearing surface during shaping. (Photo courtesy of L. A. Weaver Co., Inc.)

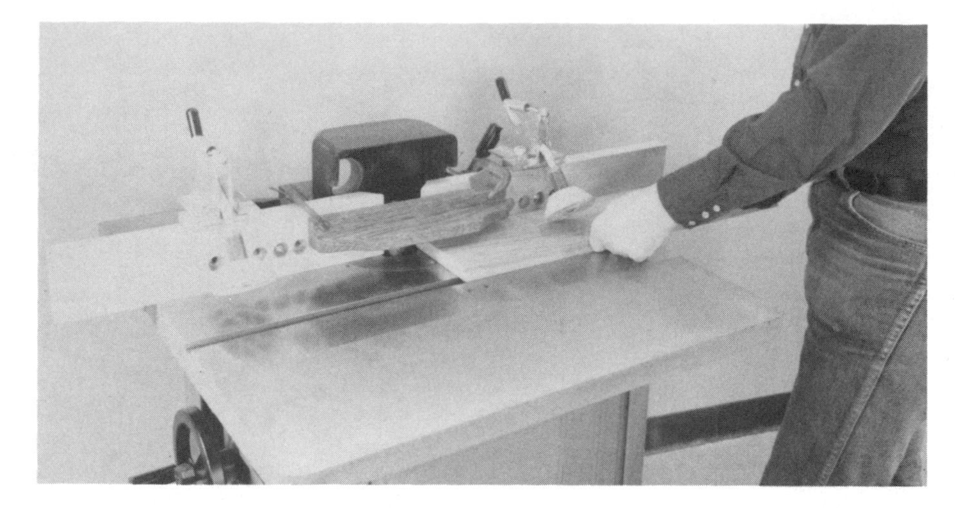

Illus. 4-34. Position yourself in the safety zone when doing any shaping. Never stand immediately behind the workpiece. You could become a target for a kickback. Keep your footing and your balance when doing any shaping operation. Avoid overreaching; this could set you up for a kickback.

examines the procedures for identifying dull cutters.

5. Guard against accidental starting. When making adjustments to the shaper, do so with the power off. It is too easy to make an adjustment that could damage the shaper or cause an accident when the power is on. Make repairs and change cutters and collars with the power disconnected (Illus. 4-32). Otherwise, a serious accident could occur.

6. Inspect your stock. Before shaping any stock, look it over (Illus. 4-33). Loose knots, twists, cupping, or rough, wet lumber can mean trouble.

Loose knots can be ejected when they contact the cutterhead. They can also be the cause of a kickback. Rough, warped, or wet lumber can also cause kickbacks.

Small parts can pose a problem as well. Machining them puts your hands too close to the cutterhead. Whenever possible, shape large pieces and hand-saw them to their correct sizes afterward.

7. Position yourself. Stand clear of the kickback zone when operating the shaper (Illus. 4-34). The kickback zone is on the thrust side of the cutter along the fence. Anything in this zone is a kickback target.

When shaping, keep sound footing and avoid overreaching. Losing your balance can lead to an accident. Make sure that you are feeding against the rotation of the cutter, and that the cutter is turning in the right direction.

8. Take light cuts. One chief cause of kickbacks is the practice of taking too heavy a cut. It is far better to take two light cuts than one heavy cut. The size of the cut depends upon the following factors: 1, the hardness of the wood; 2, the amount of stock to be removed; 3, the power of the shaper; 4, the diameter of the spindle; 5,

Illus. 4-35. Always try to make a light cut. Remember, almost twice as much stock is removed by shaping the tongue as when shaping the groove. This means that the tongue is a heavier cut than the groove. Use good judgment when setting up the shaper, and consider making light cuts on the profile.

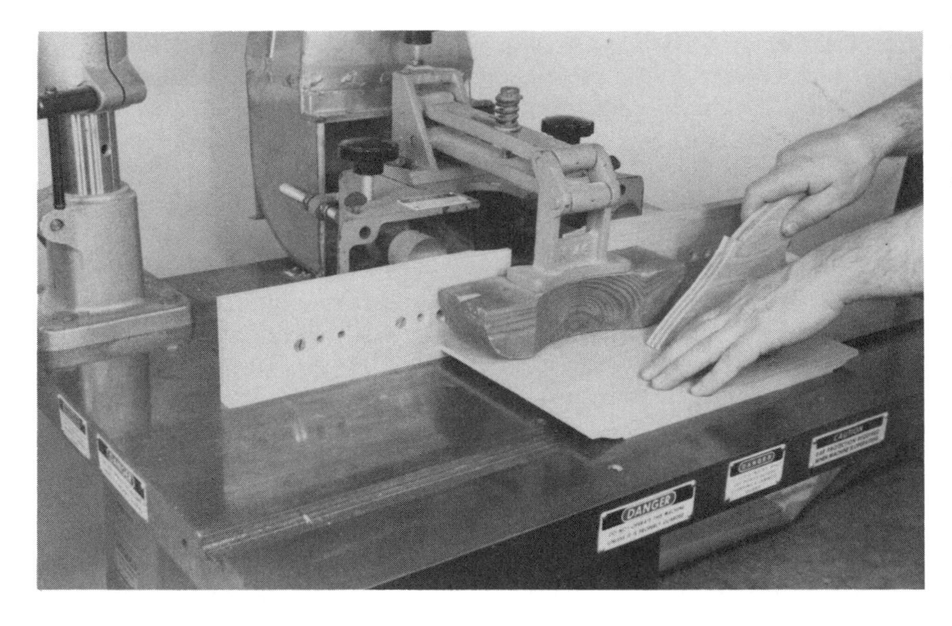

Illus. 4-36. Push blocks or push sticks can help make the operation safer. Use these with other holding devices whenever possible.

the height of the cutterhead; 6, the diameter of the cutter; and 7, the size of the workpiece.

Let's look at the tongue-and-groove joint as an example of how to determine how much stock to remove. Almost twice as much stock is removed shaping the tongue than is shaping the groove (Illus. 4-35). Therefore, the tongue-shaping cut is about twice as heavy as the groove-shaping cut. Your shaper may be able to cut the tongue or groove in a single cut. You may have to make two light cuts when shaping the tongue on very hard wood.

If the shaper is underpowered or if the cutter stack is too high, then there are likely to be cutting problems. The cutters are likely to dig into the work and cause a kickback. The larger the diameter of the cutter, the more power you need.

The diameter of the spindle can also pose a problem. Cutters should not be stacked on shapers with a ½-inch-diameter spindle. There is too much deflection in the spindle when cutters are stacked.

Experience will determine how large of a cut you can take with your shaper. While you are learning, it is best to take a light cut.

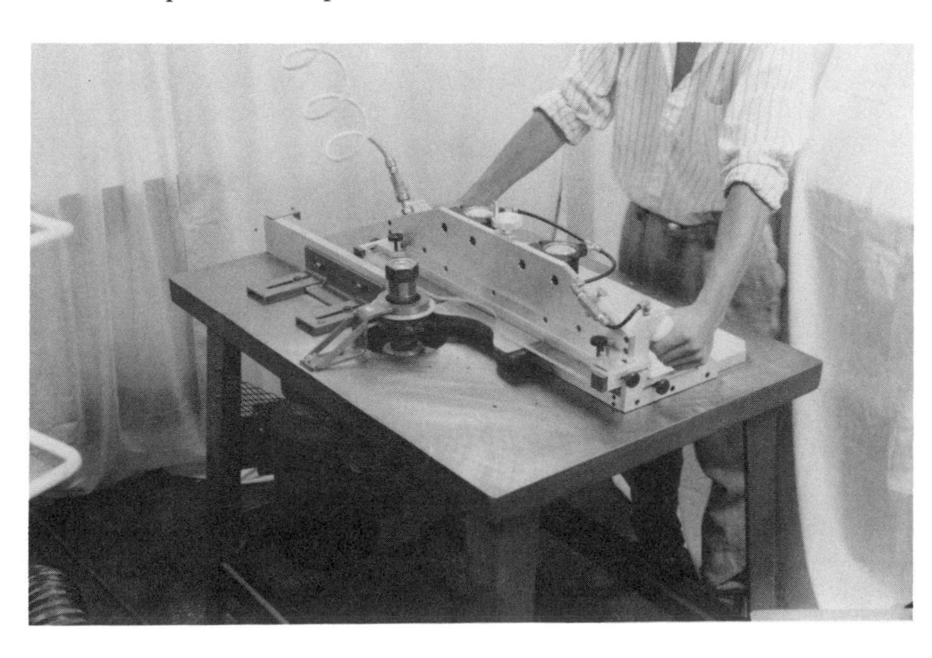

Illus. 4-37. Use jigs and fixtures to control the work and keep your hands a safe distance from the cutterhead. Jigs like the ones shown here allow you plenty of time to react in the event of a hazardous situation. (Photo courtesy of L. A. Weaver Co., Inc.)

9. Use control devices. Devices like push sticks and push blocks make handling stock safer (Illus. 4-36). These devices get in close and control the stock. Your hands are well away from the cutter, in a safe position. Keep control devices near the shaper at all times.

10. Keep your hands clear. By keeping your hands a safe distance (6 inches) from the cutterhead, you allow a margin for error. When your hands are a safe distance from the cutter, there is time to react in a hazardous situation (Illus. 4-37). Push sticks, push blocks, and hold-downs help keep your hand away from the cutterhead.

11. Check all jigs, fixtures, or other accessories. Make sure that they can be used safely, and that they are mounted securely (Illus. 4-38). Check the device with the power disconnected to be sure that it will work correctly. Never make this test with the power connected.

12. Keep fences as close to the cutter as possible. Cope them for a tight fit using a scroll saw or band saw. When the fence is close to the cutter, there is less chance of kickback, and the quality of the cut is improved. This is because the fence acts as a chip breaker.

13. When freehand shaping, be sure to use a starting pin. Begin the cut on edge grain and complete the cut on end grain. If you begin on end grain, a kickback is likely to result.

14. Avoid shaping short, thin, or narrow parts. The shaper has enough horsepower to shatter these parts or cause them to kick back. If you shape these parts, your hands will be dangerously close to the cutterhead. When in doubt, shape a larger piece. Then saw the piece from it.

15. Think about the job. When performing a new operation, think about the job before you begin. Ask yourself what would happen if you were to make the cut with the shaper set up the way it is. Questions of this nature help you identify and avoid a potentially dangerous situation.

If you have a premonition of trouble, stop. Avoid any job that you are unsure of. Try setting the job up another way, or ask some other experienced operator for his opinion.

16. Be Prepared. Read the owner's manual (and any accessory manual) before you operate the shaper. All shapers are different; make sure you understand the one you are using (Illus. 4-39).

Illus. 4-38. Whenever using any jig, fixture, fence, or any other device with the shaper, make sure that it is mounted securely and that it can be used safely for shaping. Always check its working operation with the shaper's power disconnected.

SAFETY FIRST

PLEASE POST THESE SAFETY RULES NEAR YOUR MACHINE FOR REFERENCE

SAFETY RULES for DELTA

WOOD SHAPERS and OVERARM ROUTER/SHAPER

1. **READ** the instruction manual before operating your machine.
2. **IF YOU ARE NOT** thoroughly familiar with the operation of Wood Shapers and Overarm Router/Shapers, obtain advice from your supervisor, instructor or other qualified person.
3. **REMOVE** tie, rings, watch and other jewelry, and roll up sleeves.
4. **ALWAYS** wear safety glasses or a face shield.
5. **MAKE SURE** wiring codes and recommended electrical connections are followed and that machine is properly grounded.
6. **MAKE** all adjustments with the power off.
7. **KEEP** cutters sharp and free of all rust and pitch.
8. **NEVER** run the stock between the fence and the cutter.
9. **ALWAYS** feed against the cutter rotation, as shown in Fig. A.

10. **WHEN SHAPING** with collars and starting pin, the collar MUST have sufficient bearing surface, as shown in Fig. B. Fig. C illustrates the wrong way for this operation as the collar DOES NOT have sufficient bearing surface.

11. **WHEN SHAPING** with collars and starting pin, the work must be fairly heavy in proportion to the cut being made as shown in Fig. D. UNDER NO CIRCUMSTANCES should short work of light body be shaped against the collars as shown in Fig. E.

432-01-753-0001 3-1-86

FIG. D RIGHT COLLAR CUTTER

FIG. E WRONG COLLAR CUTTER

12. **WHEN SHAPING** with collars and starting pin, the cutter should be positioned below the collar whenever possible, as shown in Fig. F.

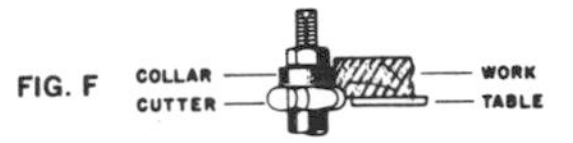

13. When the motor on the Overarm Router/Shaper is positioned ABOVE THE TABLE, the feed of the work piece is LEFT TO RIGHT, as shown in Fig. G. When the motor is positioned BELOW THE TABLE, the feed of the work piece is RIGHT TO LEFT, as shown in Fig. H.

14. **THE FENCE** should be adjusted endwise so the opening is never more than is required to clear the cutter.
15. **ALWAYS** use a miter gage and clamp attachment when edge shaping work less than 6" wide. Fence should be removed during this operation.
16. **DISCONNECT** machine from power source when making repairs.
17. **BEFORE LEAVING** the machine, make sure the work area is clean.

Illus. 4-39. Study these safety rules to be sure that you know how to use the shaper safely. When in doubt, ask someone with experience before taking a chance. If you have a premonition of danger, do not attempt the operation. (Drawing courtesy of Delta Industrial Machinery Co., Inc.)

5

MAINTENANCE, TROUBLESHOOTING, AND BUYING GUIDELINES

The shaper, like all woodworking equipment, must be maintained properly if you want to produce quality millwork and prevent injury. The troubleshooting procedures are the same, no matter which brand of shaper you use or whether it is new or used. Most new shapers that you purchase will be supplied with maintenance and troubleshooting guides.

Troubleshooting

With every operation performed on the shaper, there are routine troubleshooting techniques used. Many problems that arise in shaper operations have to do with setup procedures and safety precautions outlined by the manufacturer.

The first step in troubleshooting itself is to go through your safety procedures. The most important action to be taken is to turn off the power and disconnect the machine. Then you should check the following: the rotation of the spindle, the mounting of the cutterhead, the balance of the cutter, the fence placement and adjustment, the hold-downs, and the shaper table. Each of these factors is discussed below.

SPINDLE ROTATION

To ensure that your cutters will rotate properly, rotate the spindle in the direction you will be cutting, and check for any odd noises, binding, or stiff movement. If the spindle is belt-driven, check the belt and pulleys for alignment and looseness.

MOUNTING THE CUTTERHEAD

Check to make sure that the cutter you are using has been mounted firmly to the bottom of the spindle ring; and that enough spacer collars are being used above the cutter so that you can tighten the spindle nut firmly to prevent a free-spinning cutter.

THE BALANCE OF THE CUTTER

If the cutter is a solid-body cutter—either high-speed steel or tipped with carbide or Tantung—check with a dial indicator to ensure that the tool is concentric (Illus. 5-1). First check the cutterhead and then the spindle. Remove the cutter from the arbor and use the dial indicator on the spindle.

Check the spindle for runout using a dial indicator (Illus. 5-2). Place the indicator in a magnetic base for stability, and position it on the shaper table close to the spindle. Rotate the spindle by hand after you have unplugged the machine.

Check runout at different heights of the spindle. If the runout is above the manufacturer's recommendations or specifications, either replace the spindle (Illus. 5-3) or take it to a machine shop to correct this problem, where it will be turned on a lathe until it is concentric.

Check also for burrs on the spindle and discoloration of the spindle. This may mean the cutter, by spinning freely, has created friction

Illus. 5-1. Use a dial indicator to check the cutter to ensure that it is concentric with the spindle.

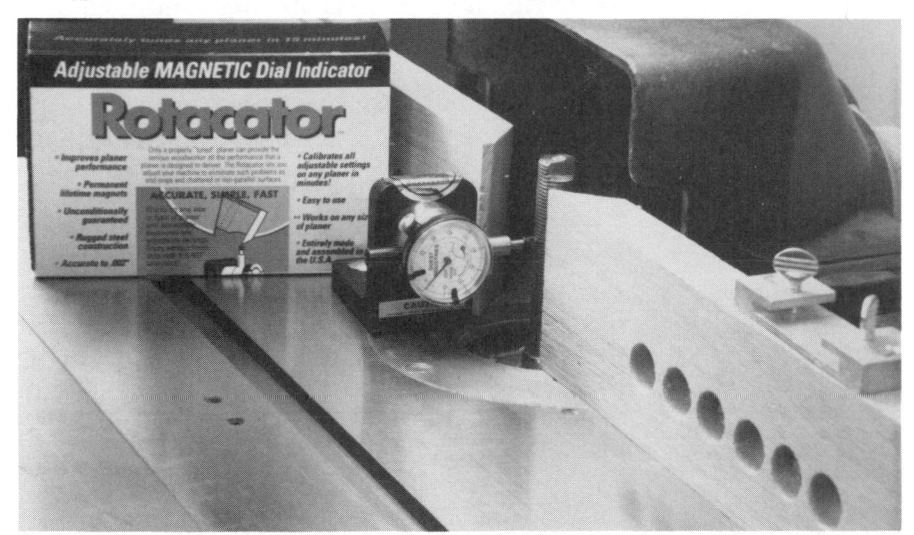

Illus. 5-2. You can also use a dial indicator to check the spindle for runout.

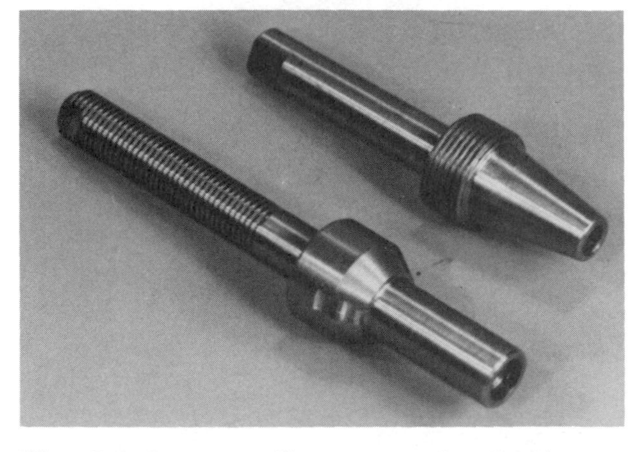

Illus. 5-3. Some spindles are interchangeable. This makes them easier to replace. Make sure that the spindle is clean before installing it.

Illus. 5-4. Always cope the fences so that they will be close to the cutter. This will reduce the size of the chips and the possibility of tearout.

with the spindle and produced enough heat to discolor the steel. This expands and contracts the spindle, possibly causing the spindle to have extreme runout.

Runout can be described as .002-.005 inch movement as the spindle is rotating; this, in turn, will increase the runout on the cutterhead. Runout will create vibration in the machine and result in an unsafe machine operation. The more vibration that occurs, the more millmarks you will receive as you shape the wood. Hand-feeding stock in this case is unsafe, and must be avoided.

FENCE PLACEMENT AND ADJUSTMENT

Check to make sure that the fence or fences, if you use a split-fence system, is bolted down securely. To determine this, consider the following points:

1. Is the infeed or outfeed fence touching the cutter?
2. Is the fence close enough to the cutter to reduce chip size and create a safe operation for hand- or power-feeding (Illus. 5-4)?
3. Is the fence perpendicular to the table (Illus. 5-5)?
4. Are you taking too large a cut with one pass?

By adjusting the fence in and out, you can reduce the depth of cut.

HOLD-DOWNS

Ask yourself the following questions about hold-downs:

1. Are the hold-downs used preventing the cutter from rotating freely?
2. Have they been placed properly to create a safe operation and hold the workpiece down and against the fence securely (Illus. 5-6)?
3. Could the tension on the hold-downs be increased to increase control over the workpiece?
4. Are the hold-downs clear of the cutter's path (Illus. 5-7)?
5. Will the hold-downs act as a barrier between the cutter and the operator?

SHAPER TABLE

The shaper table should be kept clean and free of oxidation. Auto-body rubbing compound can be used to remove oxidation. Use a good grade of paste wax (Illus. 5-8) to protect the table. Apply the paste wax periodically to keep the table from oxidizing and to make stock-feeding easier (Illus. 5-9).

Illus. 5-5. Check the fences to be sure that they are perpendicular to the table.

Illus. 5-6. The featherboard will hold stock against the fence. It also acts as a barrier between the operator and the cutter.

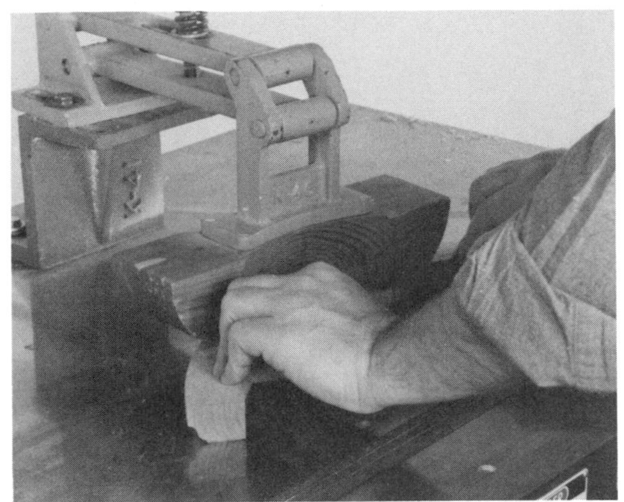

Illus. 5-7. Hold-downs should always be clear of the cutter's path.

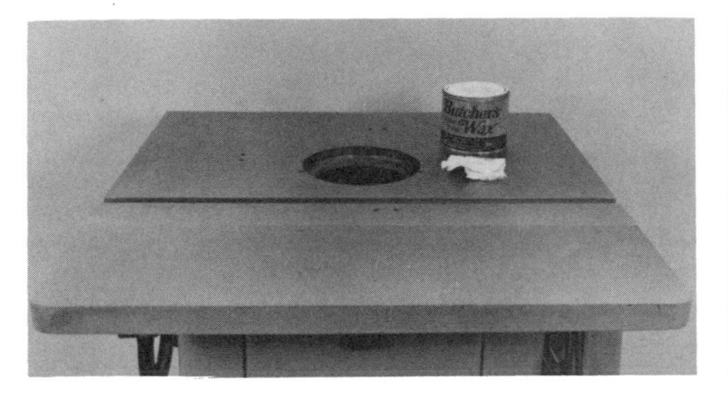

Illus. 5-8. Use a good grade of paste wax to protect the shaper table.

Illus. 5-9. The paste wax will also make it easier to feed stock.

Power Feeders

Check the power feed to make sure it is aligned with the table and fence. The rollers on the feeder should be a safe distance away from the cutter. Each setup is different, and the feeder needs to be adjusted accordingly.

If you are using the feeder horizontally, make sure that the feed wheels are at least 1/4 inch below the workpiece for a safe operation. Also, place the feed unit so that it tapers in towards the fence at least 1/4 inch per foot.

When you are using the feeder in a vertical position, the pressure on the workpiece is towards the fence, and the feeder tapers towards the table. Make sure that the fence is locked securely at the desired setting. The feed pressure of the power feeder could move the fence.

After you have gone through these steps, it is a good idea to run a sample board through your setup and look at the result. While it is running through the shaper setup, listen for or try to feel any vibration. By placing a hand on the base of the machine, you may be able to pick up unnecessary vibration. Do not touch any moving parts or examine internal components while the machine is running. Make sure that you check the pulleys and belts of your machine before you turn the machine on.

Check the wheels on the power feeder periodically for wear. Replace the wheels when they become damaged. When working with woods containing excess sap, you will have to clean the wheels periodically to maintain traction. Be sure that the solvent you use does not damage the wheels. Be sure to grease any fittings on the power feeder. Consult your owner's manual for specifics.

Maintenance Procedures

The following procedures are routine maintenance procedures. Follow each and every one before any operation.

1. *Check all bolts, screws, and locking devices used to hold the fence in place.* Replace any threaded fasteners that have been stripped through constant use. Replace any bolts or nuts with worn heads. These bolts or nuts can be difficult to tighten, and allow wrenches to slip.

2. *Check the surface of the table for burrs and rust.* Clean, file, and buff out these irregularities to prevent excess friction. A coat of paraffin wax will help keep the machine surface properly clean.

3. *Check all the table inserts to ensure that they are properly seated around the table opening.* Check for burrs and pitch. Clean them with a good solvent and apply a coat of wax.

4. *Make sure that the guards and hold-downs are working properly.* Make sure that they have the correct tension. If they are made of wood or plastic, make sure they have no cracks or sharp edges. Make new hold-downs and guards if possible, or check the manufacturer's parts list to see if you can order them. Preventing any kind of injury is well worth the energy it takes to design and build a better guard or hold-down.

5. *Grease any fittings the manufacturer recommends greasing.* This will prolong the life of your shaper. This, however, will not bring your machine back to life.

6. *Check the belts and pulleys for adjustment of tension.* Replace belts that look worn or frayed. Pulleys can also wear if too much pressure is applied to them. The friction between the belt and pulley can reduce the life of the pulley. Check the fit of the pulleys to the motor arbor and spindle. Stopping and starting the machine operation can, in time, loosen and wear down these shafts. Replace pulleys and woodruff keys.

7. *Check the spindle for burrs and scratches.* By keeping the spindle nut tight and placing your cutters onto the spindle carefully, you can prevent having to replace the spindle, which can be costly. You can use a fine-tooth file to take a burr off the spindle, but be careful; removing too much material will definitely reduce the effective diameter of the spindle and create more runout.

8. *Check the mitre gauge slot for wear and burrs.* Clean it with a good solvent and repair the slot by filing the burrs carefully. Apply some paraffin wax to reduce friction.

9. *Check the mitre gauge for loose or broken parts.* Replace any parts that look worn, such as a threaded locking nut or pivoting arm attaching bolt.

10. *Check all electrical connections.* Make sure that there are no loose or broken wires. With the power disconnected, use compressed air or a soft brush to clean the connecting terminals. Dust will create electrical shorts. It is important to check this frequently.

11. *Check all the moving parts on the spindle carriage or assembly for wear.* Some shapers have the tension adjustments right on the assembly to hold it in place during the cutting operation. For example, the adjustment may consist of tightening two or three gib screws or adjusting a tension bar with one bolt. Lubricate these moving parts periodically according to the manufacturer's recommendations.

SPECIALTY MAINTENANCE AND TUNE-UP

When you purchase a new or used shaper, you should level it and make sure that the spindle is square with the machine top. Use wood shims under the machine base to level the top; they will dampen vibration, which will result in better finished work.

Ideally, you should mount your shaper to the floor. This will prevent injuries that occur when the shaper walks or tips over during the shaping operation. If you cannot fasten the shaper to the floor, clamp the base of the machine to a workbench to give it additional support and weight.

Some rolling bases can be used with lighter cabinetmaker's shapers. When they are used,

Illus. 5-10. This rolling base allows the shaper to be moved around the shop or out of the way.

Illus. 5-11. Always lock the wheels before shaping. Always make sure that the shaper is not moving before you begin.

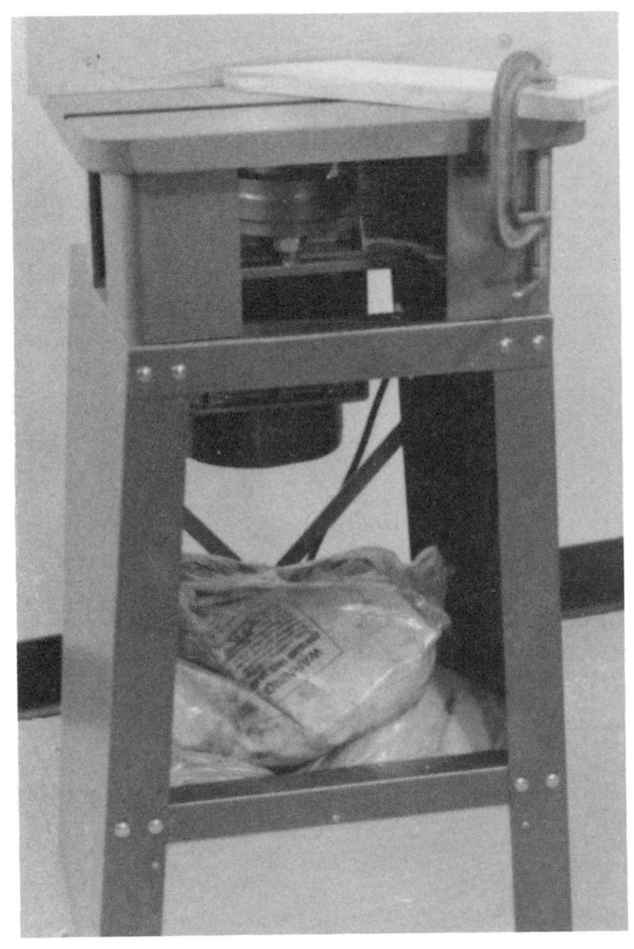

Illus. 5-12. The weight of the sand in the base of this shaper will keep it from deflecting even during the heaviest cut.

the machine can be moved out of the way or into a new position for long or curved work (Illus. 5-10 and 5-11). Be sure that the power feeder is kept over the table if you plan to move the shaper with the power feeder attached. If you move the power feeder out of the way, the weight of the power feeder may be sufficient to topple any shaper that is not attached to the floor.

Some light-duty shapers are actually top-heavy. This could cause the shaper to topple over in the midst of a heavy cut. You can eliminate this problem by increasing the weight of the shaper base. The sandbags placed in the base of the shaper shown in Illus. 5-12 ensure that it will not deflect or topple even during the heaviest cut. This practice is effective in shops where the shaper cannot be secured to the floor.

Check that the spindle is square with the table by placing a good machinist's square on a clean surface. Touch the square to the spindle. Use either paper or metal shim stock to check the full length of the spindle. If you detect a problem when using the square, loosen the bolts securing the shaper table and shim between the table and the base. Make adjustments until the table is square with the spindle.

Check the fence system on the shaper by holding a square perpendicular to the tabletop. You may have to shim to correct any irregularities. In some cases, the fences can be held against an edge-sanding machine and squared relative to the bottom of the fence carriage.

Getting the Best Cut from Your Shaper

To achieve the best possible cut from your shaper, you must consider the material which is to be cut. Also, you should consider how many lineal feet of stock or edge will be run through your shaper. You can select a cutter appropriate for the length of run and material to be cut, and that will last or stay sharp throughout the entire run. Once you have selected the cutter, adjust the spindle speed so that it correlates with the largest diameter of the cutter. Knowing the rpm and diameter of the cutter, you can determine what speed of feed should be used. The life of your shaper and your cutters will depend on your ability to select the correct spindle speed, feed speed, and type of cutter (always sharp) for the material you will be shaping.

A good finish cut from your shaper will also depend on the procedures you use to set up your fence and hold-down system (Illus. 5-13). Keep your fence as close as possible around the cutter being used. This prevents injury and reduces chip size in all shaping conditions. Cope your fences—both the infeed and outfeed side of the fence—as close as possible to the cutter profile. Do this with a scroll saw or band saw.

Always wax your fence and table to prevent friction. A consistent feed will result in a uniform cut.

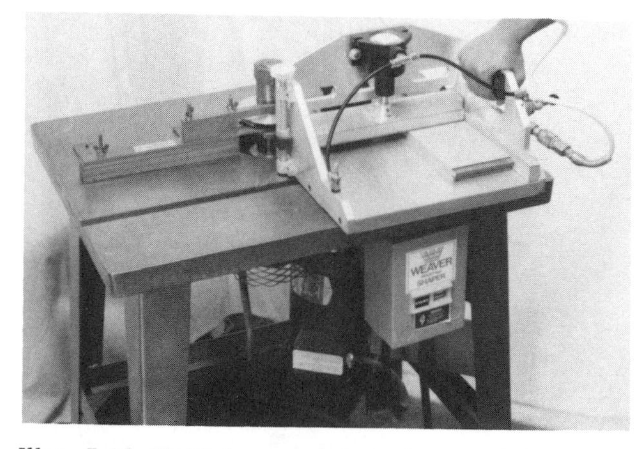

Illus. 5-13. Hold-downs and stock-control devices can affect the quality of the cut you obtain from any setup. (Photo courtesy of L. A. Weaver)

Be aware of changes in the finish cut while you are shaping. This gives you the opportunity to change or hone the cutter, or change the rpm and feed speed, before you've run all the stock incorrectly.

Make sure that the power is disconnected before making any changes. Also, make sure that the changes being made are safe. Take notes as you go. Save a sample of the moulding with the correct spindle speed, feed speed and other pertinent information written on it. This sample will help you make the next setup in much less time.

Buying a Shaper

There are many sizes and types of shapers available on the market today. The manufacturers of shapers continually try to improve the safety features and operating capabilities of their machines. Dust collection, the fence system, and a sliding table are some of the shaper features the woodworker is concerned with. Many manufacturers have greatly improved these features. Cutters have also been improved. When buying a shaper, also look at the many cutters that are available. The shaper that you buy will only be as good as the cutters you are able to use with it.

Following is a checklist of shaper features you should examine before buying a shaper. This list will help you find the shaper that will best suit your needs. Parts of this list apply to both new and used shapers. You may want to add to this list, depending on your needs.

Shaper Buying Checklist

A. SIZE OF THE SHAPER

1. Can the shaper top accommodate the shaping of the wood projects you are presently producing?
   a. Does the shaper top provide the space for a power-feed unit?
2. Can the shaper hold an auxiliary tabletop mounted over the manufactured top without

the efficiency of the original machine being reduced?

3. Are table extensions available for the shaper?
   a. Can homemade table extensions be adapted to the shaper top?
4. Does the height of the shaper seem comfortable when you are freehand shaping?
   a. Can the shaper be adjusted for height, and also levelled for stability?
5. Does the weight of the shaper provide the stability you need for freehand shaping?
   a. Will the weight and size of the shaper accommodate a power-feed unit?
   b. If the center of gravity on the shaper is too high or off center, can the shaper be bolted to the floor?
   c. Can the shaper be moved around the shop or at a job site?

B. MOTOR SIZE AND ELECTRICAL PARTS

1. Does the motor horsepower and amperage provide the power for:
   a. shaping continuous raised panels (minimum 3 horsepower)
   b. shaping continuous mouldings
   c. continuous freehand shaping
2. What are the voltage requirements of your shop?
   a. Light-to-medium duty single-phase shapers that are either 110 or 220 volts have 1 horsepower to 5 horsepower motors
   b. Heavy- or production-duty three-phase shapers have 3 horsepower to 10 horsepower plus motors
3. Should the shaper have reversing switch capabilities?
   a. In freehand shaping, is there a need to reverse spindle rotation because of the grain direction of the wood being shaped?
4. Can the machine be disconnected easily as a safety precaution?

C. RPM REQUIREMENTS

1. What diameter cutters will be used?
2. Custom woodworking requires many different diameter cutters; thus rpm from 4,000 to 10,000 would be necessary. Smaller cabinet shop shapers use smaller cutters and have two speeds between 5,000 and 10,000 rpm.

D. SPINDLE-SIZE REQUIREMENTS

1. What size jobs are being produced?
   a. For continuous duty, a solid spindle will last longer with less maintenance
   b. Removable spindles can be helpful economically when you are purchasing cutters. Smaller-diameter cutters cost less. For a light-duty shaper, buy a ½–¾-inch spindle. For a heavy-duty shaper, buy a ¾–1¼-inch spindle.
   c. Tilting spindles provide unlimited cutter profiles, a useful option to consider when you are examining the price of shaper cutters.
   d. Can the length of the spindle under the nut provide you with enough space to stack cutters and cut setup time?
   e. What size cutters are most important to your present and future needs in woodworking?

E. ACCESSORIES

1. Is it necessary to have a split or one-piece fence system?
   a. Can the fence be adjusted easily and with accuracy?
   b. Does the fence system provide for dust collection?
   c. Does the fence system provide for removable and adjustable wood or composite platens that can be coped for any of your shaper cutters?
   d. Can the fence system be completely removed for freehand shaping?
2. Is there a mitre slot and gauge provided with the shaper?
3. Is a sliding table necessary?
   a. Are you cope-and-stick cutting, tenoning for mortises, or using special locking shaper cutters?
   b. Does the sliding table come close enough to the spindle and cutters to be accurate?
   c. How much deflection does the table have?
4. Are hold-downs provided for moving stock safely through the cutting circle?
   a. Can the hold-downs be used when you are freehand shaping?
   b. Are starting pins for freehand shaping provided? Are there several positions for the starting pins?

Cost is also a factor when you are selecting a shaper, but never let it override the safety factors. The checklist for buying a new or used shaper also applies when you are buying cutters for the shaper. It would be a mistake to buy cutters without considering your needs for a shaper, and, in turn, you should consider all the cutters available for standard shapers with a bore diameter that meets your shop needs.

Besides the factors just explored, consider these additional factors when buying a used shaper:

1. Does the spindle appear to be worn or scarred?
2. Is the drive belt worn?
3. Can you feel side-to-side and up-and-down movement of the spindle?
4. Does the shaper have its original motor?
5. Has the shaper had any visible repairs done to it, such as welds or bolts holding machined parts together?
6. Has the top been altered for a custom operation?
7. Is the spindle-height adjustment still doing its job?
8. Is there a fence with the shaper?
9. Will its electrical output meet the needs of your shop?
10. Has the electrical system been altered?
11. Consider the overall appearance of the shaper. Has the shaper been taken care of?

Many used shapers are sold with shaper cutters and knives which may not have ever been used on the machine you are purchasing. Only consider buying solid-body, high-speed-steel or carbide-tipped used cutters. Take these cutters to a professional grinder and have them checked for balance, bore concentricity, or scoring, sharpening, and tip replacement. Do not sleeve down 1¼-inch-bore production cutters to ½-inch spindle shapers. Call or write to the manufacturers of both the machine which is for sale and the cutters available for complete specifications on those tools. Manufacturers are always happy to help supply any information relating to their equipment. Serial numbers and model numbers are important, and sometimes can even date a machine.

The most important thing you should do when buying a shaper, either new or used, is evaluate your needs, both present and future. Always be aware of the safety features for each shaper you are considering. The shaper can provide unlimited possibilities in the design and construction of your future woodworking projects.

6
BASIC SHAPER OPERATIONS

Installing a Cutter for Edge Cutting

Before working on the shaper, disconnect the power. Unplug the machine or disconnect it at the circuit box. Lock the power box to keep others from turning it on without your consent. Clear away the fences and chip hood for greater access to the spindle.

To install the cutter, first remove the spindle nuts and any existing cutters or spacer collars (Illus. 6-1 and 6-2). Some woodworkers feel that the cutter should go in the middle of the spindle or slightly lower. This allows the spindle to be raised or lowered for additional cuts. Other woodworkers feel that the cutter should be closer to the top of the spindle. This minimizes the amount of spindle in the way of the guard or dust collector. If the arbor diameter is small, it is best to keep the cutter low on the arbor. This will minimize arbor deflection.

Collect the appropriate collars and inserts (Illus. 6-3). Mount the spacer collars on the spindle first (Illus. 6-4). This will position the cutter at the correct height.

If you are using a live-bearing spacer collar, install a shim over it and then position the cutter on the spindle. Be sure to consider spindle

Illus. 6-1. After removing the spindle nuts, remove the spacers and arbor. Maple blocks can sometimes be used to lift the cutters and spacer collars from the spindle. Withdraw the spindle as the maple blocks hold the spacer collar and cutters.

Illus. 6-2. If you use maple blocks to withdraw the spindle from the spacer collars or cutters, make sure that the carbide tips of the cutters do not touch the maple blocks. This could damage them.

Illus. 6-3. Before setting up the shaper, make sure that you have the appropriate table inserts, spacer collars, cutters, nuts, etc., available for use.

Illus. 6-4. Spacer collars are usually mounted on the spindle first. This positions the cutter at the correct height.

Illus. 6-5. Be sure to consider the rotation of the spindle before mounting the cutterhead. Some cutters are designed for one direction of cut only.

Illus. 6-6. When using assembled cutters, make sure that the cutters have been installed correctly and securely, and that they have the same profile.

Illus. 6-7. An extra spacer collar is used above this cutter so that the spindle nut will hold the cutter securely in position.

Illus. 6-8. Some shapers use a bearing washer. If one is used on your shaper, install it before mounting the spindle nut.

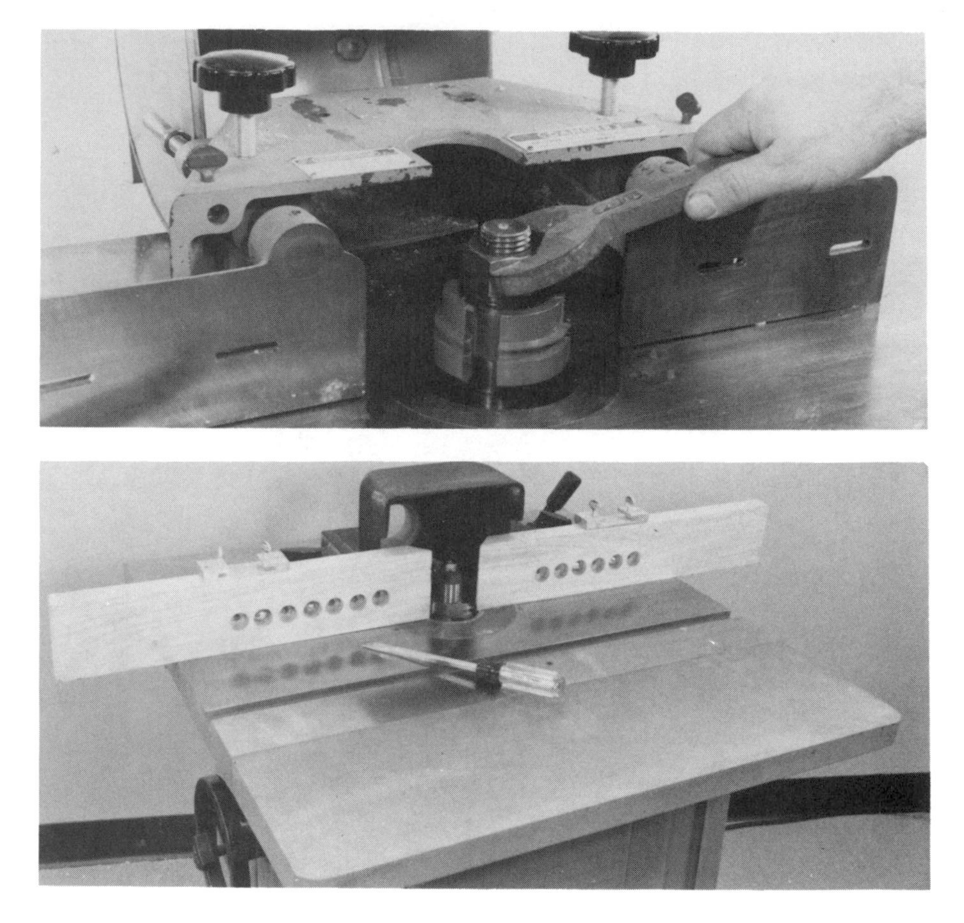

Illus. 6-9. Be sure to tighten the spindle nut securely. Some machines have a spindle lock that tightens the nut; others require two wrenches. In either case, make sure that the spindle nut is locked securely.

Illus. 6-10. After mounting the cutter correctly, mount the fences and adjust them for the cut.

rotation before mounting the cutterhead (Illus. 6-5).

Some cutters are designed for only one direction of cut. Be sure to install those cutters correctly. When using assembled cutters, make sure that the cutters are installed correctly and securely (Illus. 6-6). It is very easy to mix the cutters up. Also, make sure that all cutters in the cutterhead are identical in shape.

At least one spacer collar should go over the cutter. Be sure to add a shim if you are using a ball-bearing collar. Use more spacers until the top one is over the threaded portion of the spindle (Illus. 6-7). This allows the spindle nuts to clamp the spacers and cutter securely.

Mount the bearing washer if one is used, and then mount the spindle nut (Illus. 6-8). Tighten the spindle nut securely (Illus. 6-9).

On some shapers, you lock the spindle to tighten the spindle nut. On other machines, two wrenches are used. If you are using two wrenches, after you have tightened the first nut, tighten the second one against it. (*Note:* Not all machines use two spindle nuts, however; reversible shapers use two spindle nuts or a special washer. This eliminates the chance of loosening the nut when reversing spindle rotation. If your shaper uses two spindle nuts, be sure to use both of them. Be sure that the threads of both nuts are fully engaged with the spindle.)

Check the table insert. The table insert reduces the size of the arbor hole. The insert should reduce the arbor hole to a diameter slightly larger than the cutter. It gives the work additional support during the shaping operation.

Attach the insert securely to the table. This is important. It is possible that the air velocity of the spinning cutter can actually lift a table insert out of the table. This could lead to a serious accident. Also, make sure that the cutter will not touch the table insert.

Now, adjust the cutter relative to the cut that will be made in the workpiece. To raise or lower the cutter, first unlock the elevating wheel. Next,

Illus. 6-11. In many cases, it is helpful to use the stock you plan to shape to lay out the position of the cutter and the fences.

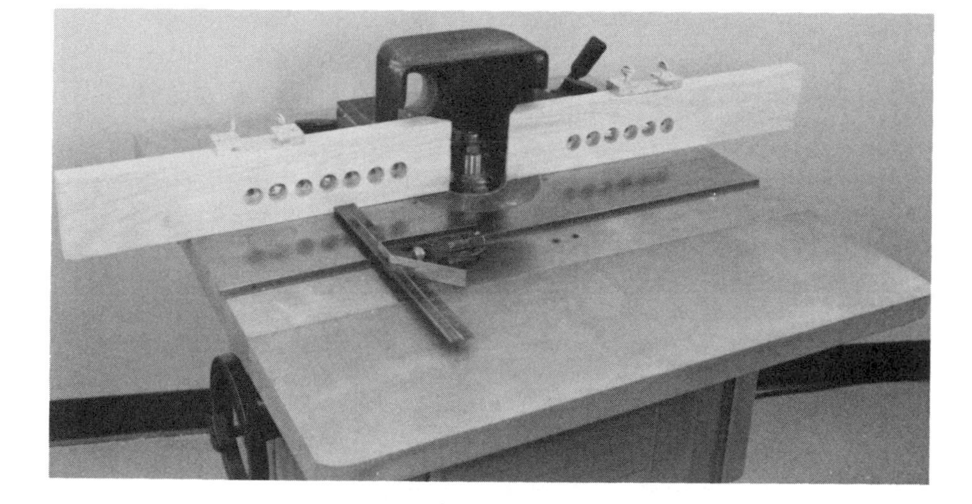

Illus. 6-12. If you are going to use a mitre gauge in the shaping operation, then you must position both fences the same distance from the mitre slot before setting up the cut. Do this using a combination square.

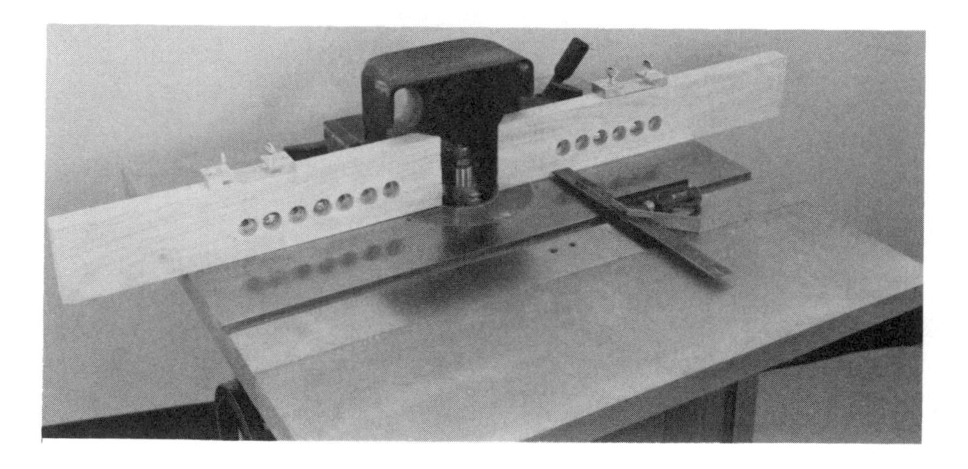

Illus. 6-13. Make sure that both the infeed and outfeed fences are equidistant from the mitre slot.

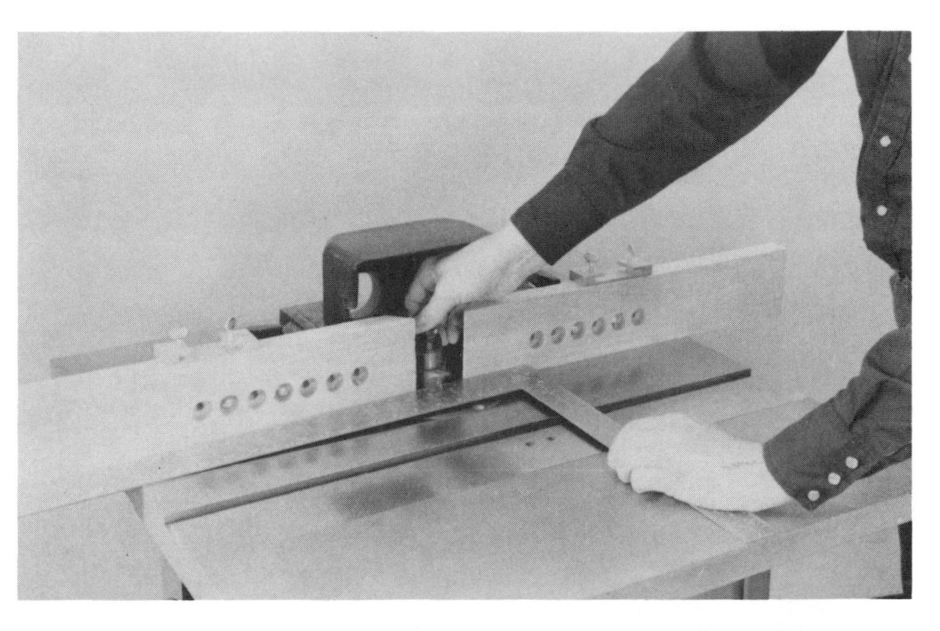

Illus. 6-14. Adjust the outfeed fence so that it is even with the deepest part of the cut made with the cutter. In some cases, the first cut may only be half that depth, so a later cut can be taken.

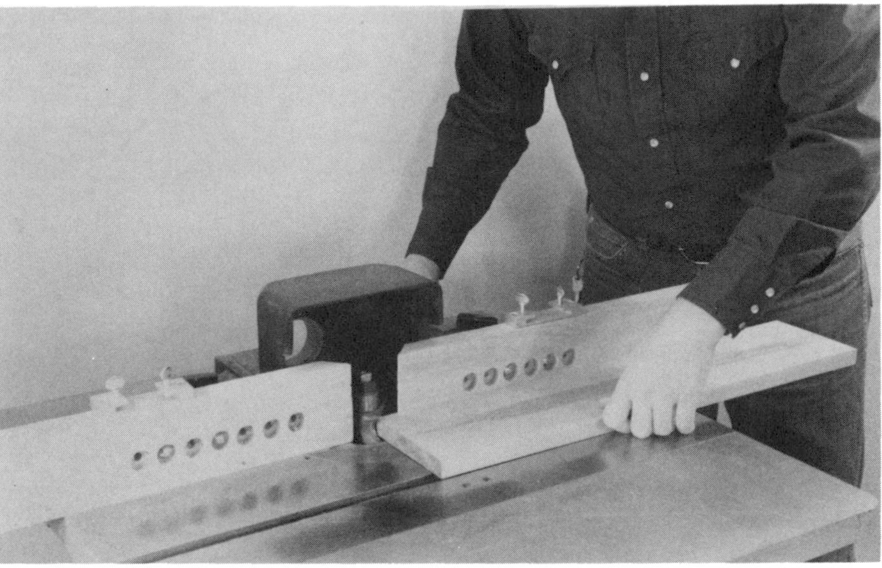

Illus. 6-15. Adjust the infeed fence for the depth of cut by moving it rearward relative to the cutter and the outfeed fence.

Illus. 6-16. In some cases a ring-shaped guard is mounted to the spindle above the cutter to prevent the operator from coming into contact with the cutter or spindle. In this case, you have to cut the fences away to make the setup.

Illus. 6-17. Here, a wooden barrier has been clamped to the fences to act as a guard above the workpiece. This guard protects you from coming into contact with the cutter. Note that because offset fences are being used, a shim has been placed between the barrier guard and the infeed fence.

Illus. 6-18. This barrier guard above the cutter also acts as a hold-down. When you tighten the nut with a wrench, you increase the tension on the hold-down.

crank the elevating wheel to position the cutter. Use the layout stock as a guide to determine where the cutter must go. Once the cutter is positioned, lock the elevating wheel.

Now, mount the fences (Illus. 6-10) and adjust them. Use the layout stock to position the fences (Illus. 6-11). Move the fences as close to the cutter as possible. The closer the fence is to the cutter's edge, the safer it is to use.

If the mitre gauge will be used in the shaping operation, you will have to align the fences parallel to the mitre gauge (Illus. 6-12 and 6-13). This process must be performed to ensure accurate shaping of end grain.

If the shaper is set up to remove the entire edge of the workpiece, the fences will remain in a straight line or parallel with the spindle. If the entire edge will be removed, the fences must be offset. Adjust the outfeed fence so that it is even with the deepest part of the cut (Illus. 6-14). Adjust the depth of cut by moving the infeed fence rearward relative to the cutter and the outfeed fence (Illus. 6-15). Once the fences have been adjusted, lock them in position.

Now, position the guard over the cutter (Illus. 6-16). Ring-shaped guards can be mounted on the side of the shaper table. Other guards are mounted on the fence (Illus. 6-17) or chip hood

Illus. 6-19. In some cases, the barrier guard is a cylindrical device that goes above the cutter. This device also prevents the operator from coming into contact with the cutter. A dust-collection hose can also be inserted into the top of this cylinder to remove chips from the top of the cutting operation.

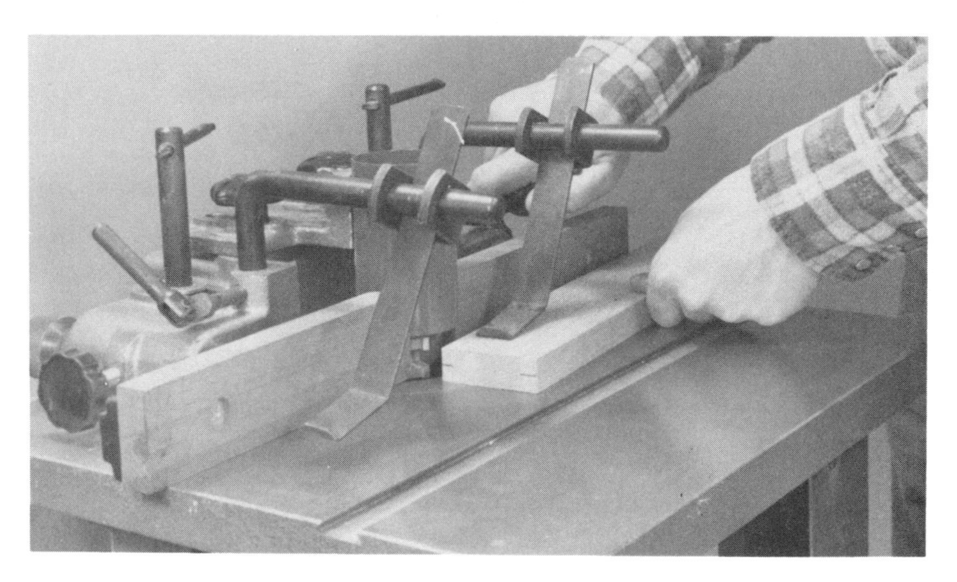

Illus. 6-20. Spring-loaded hold-downs are used in this setup to keep the stock firmly against the table. This does not hold the stock firmly against the fence, however. A side thrust from the operator's hands will do that.

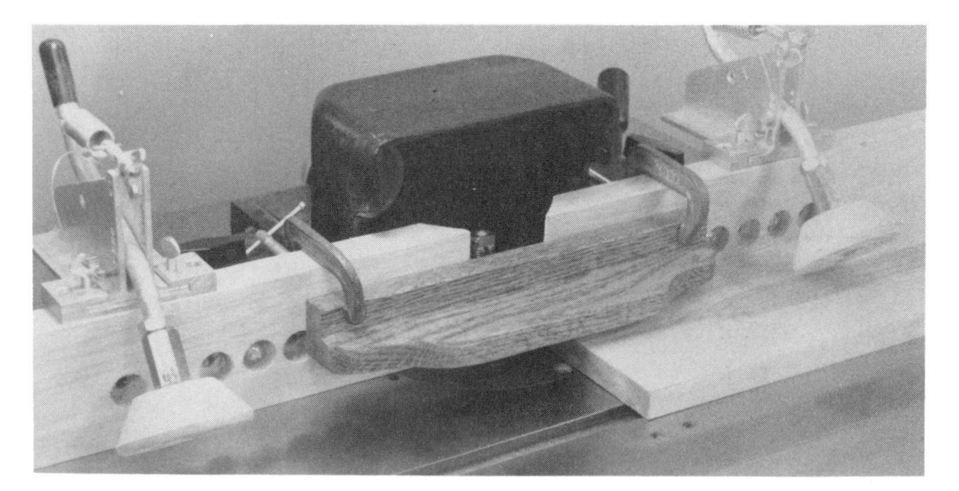

Illus. 6-21. These Shophelper wheels work very well as a holding device. They hold the stock in towards the fence and down on the table. This particular set of wheels turns only in the direction of feed. There are others that turn both ways. The difference is indicated by the color of the wheels: yellow, right to left feed; orange, left to right feed; and green, both directional feeds.

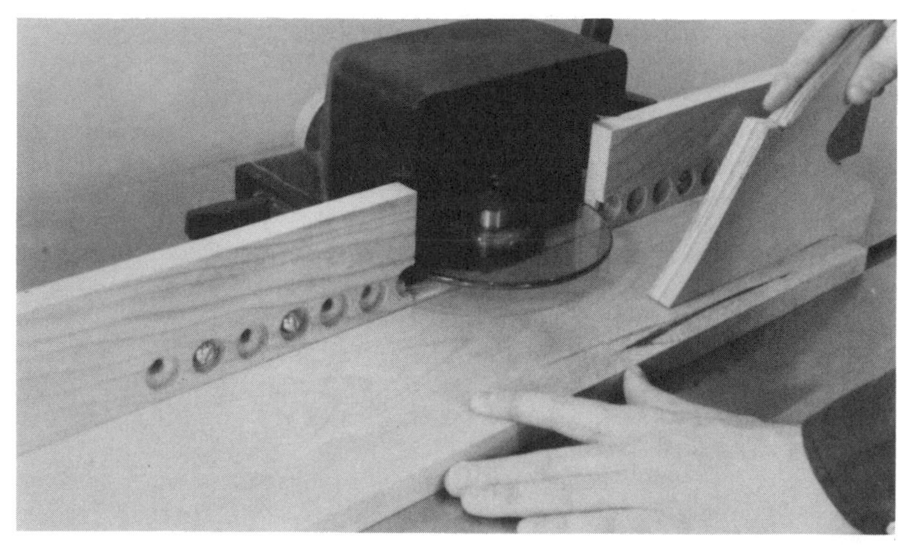

Illus. 6-22. When shaping, keep the stock firmly against the fence and on the table. Use a push stick whenever possible. Stand to the side of your work to avoid a kickback. Always wear proper eye protection when shaping.

(Illus. 6-18 and 6-19). Consult your owner's manual to make sure you have mounted the guard correctly.

Using your setup stock, adjust the shaper hold-downs to control the work (Illus. 6-20). Remember, commercial and shop-made hold-downs should be used in many setups. They make the shaper safer and easier to control (Illus. 6-21). Clamp or mount the hold-downs securely and check them occasionally while you work.

If you are using a dust collector, hook it up now. Clamp the collector hood in position or attach the collection hose to the chip collector on your shaper. Avoid kinks or sharp bends in the hose. This will reduce the dust collector's efficiency.

Before you begin shaping, check the setup. Look over all the adjustments that can be made on the shaper. Make sure that all adjustments have been made correctly, and that everything is locked securely. A loose clamp or spindle nut could mean trouble in the middle of a cut.

Check the setup before shaping. As a final check, turn the cutter over by hand to make sure it will not contact the fence, guard, or chip collector.

Restore the power to the shaper and turn it on. Make sure that the cutter is turning in the right direction. Feed the stock *against* the face of the cutter. Do not force the stock. Cutting should be smooth and even. If the stock tends to kick, or there is a great deal of chatter, the cut is too heavy. Light cuts work best. It is better to take two light cuts than one heavy one.

Inspect your stock as you work. Try to feed stock with the grain. This will improve the quality of the cut. Avoid knots and warped or twisted lumber. These stock defects can contribute to kickbacks and other hazards. It is far better to discard defective wood than to damage a cutter or spindle. Remember, the shaped portions of your project should be the most decorative. Defective wood could ruin the overall appearance of your project.

Keep the stock firmly against the fence and on the table. When possible, use a push stick to guide the stock (Illus. 6-22). Stand to the side of the workpiece. This will keep you out of the kickback zone. Remember to wear the proper eye protection while shaping.

Changing a Cutter

After a shaper cutter has been mounted and bolted to the spindle, it may be difficult to remove. This is because the tolerances between the spindle and the cutter are usually very close.

To remove a cutter, first disconnect the power to the shaper. Then move the fences and chip hood out of the way, and remove the spindle nuts.

To remove the cutter, raise it until the bottom of the cutter is well above the table. Be sure to release the lock before turning the wheel. Now you can position a small block of very hard wood under both sides of the cutter or spacer collars (Illus. 6-1). Keep the blocks away from the carbide tips on the cutter (Illus. 6-2). Now, slowly lower the spindle. Turn the handwheel while watching the cutter. The cutter should begin to lift on the spindle. Once the cutter reaches the threaded portion of the spindle, it should lift off easily. This is because the threaded portion is slightly smaller in diameter.

Most cutters come off easily, but they occasionally become stuck on the arbor. Heat and pitch buildup are the reasons for this.

The spacers and shims should lift off without difficulty after you have removed the cutter. If not, the above procedure may be used to remove them. Some spacers may not need to be removed if a similar cutter will be mounted.

Now, mount the cutter that you want to use for the next operation. Use the procedure discussed on pages 119–126. Be sure to check everything that should be checked before restoring power to the shaper.

Shaping Edge and End Grain

When a rectangular piece of wood is being shaped, two of the shaped surfaces will be edge grain, and the other two will be end grain. Generally, shaping begins on end grain and progresses around the workpiece. The last surface shaped would be on edge grain. By shaping the edges in this sequence, you will improve the quality of the work.

End grain tends to tear out when it is cut. This is because of the grain orientation. As the edge grain is shaped, the torn end grain is cut away.

When end grain is being shaped, the cut is usually controlled by the fence and mitre gauge. The end grain of the workpiece is butted to the fence. The edge grain is butted to the mitre gauge. The stock can be clamped or held against the mitre gauge. The work is then fed or guided into the cutter. The workpiece should be fed against the rotation of the cutter.

For safe shaping when freehand cutting end grain, make sure that the stock is at least 12 inches wide. Do not shape the end grain of narrower pieces without clamps or jigs.

In some cases, a wooden auxiliary face is attached to the mitre gauge. This provides support for the end grain as the cut is completed and minimizes end grain tearout.

In other cases, a rolling table is used to control stock for end-grain shaping. The rolling table may have a clamping device for increased control of the workpiece. A rolling table makes it much easier to handle large pieces. The work is well supported through the entire cut. Large pieces can be very difficult to control with a small mitre gauge.

Edge-grain shaping is usually controlled by the fence. Stock can be held against the fence with featherboards, Shophelper™ wheels, spring hold-downs, or push sticks. These devices can also be used to hold stock down on the table. Edge-grain shaping generally presents fewer problems than end-grain shaping.

Shaping Away an Entire Edge

In some shaping operations, the entire edge is removed by the shaper cutter. This occurs when the edge of the work is jointed by a straight shaper cutter or when special mouldings are cut.

When the entire edge is being shaped away, the fences must be offset to control the work on the outfeed side of the shaper. Adjust the outfeed side of the fence so that it is in the same plane as the cutting edge of the cutter. When the cutter has an irregular profile, adjust the fence to the same plane as the part of the cutting edge closest to the spindle. Remember to keep the fences as close as possible to the cutter. If two cuts are required, you may have to adjust the fence twice.

To remove the entire edge of the piece, position the infeed fence so that it is slightly behind the cutter's arc (Illus. 6-23). Position the outfeed fence so that it is even with the cutting arc (Illus. 6-24).

Feed the stock into the cutter to remove the rough edge (Illus. 6-25). The straight cutter

Illus. 6-23. Position the infeed fence for the depth of cut. This is usually no more than 1/8 inch.

Illus. 6-24. When jointing stock, position the outfeed fence so that it is even with the cutter's arc. The cutter used to joint stock is usually a straight cutter.

Illus. 6-25. Position the rough edge against the fence and shape the stock. Keep the stock against the fence with featherboards if possible.

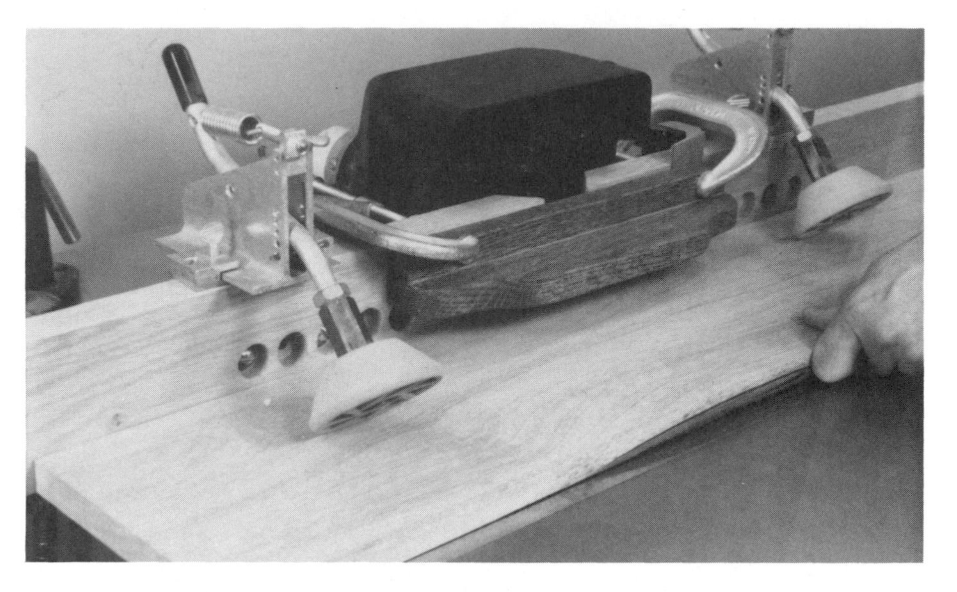

Illus. 6-26. Position a barrier guard over the cutter to prevent you from coming into contact with the cutter. Do not force the workpiece too quickly, and watch the direction of grain in your pieces. The cutter should comb the grain towards the infeed fence.

Illus. 6-27. To use the power feeder when jointing a piece, first set up a straight cutter at the appropriate height for the jointing operation.

used for this operation has to be taller than the thickness of the stock.

Position a barrier guard over the cutter to eliminate any possibility that you will come into contact with the cutter (Illus. 6-23D). Feed stock carefully into the cutter at moderate speed (Illus. 6-26). Pay careful attention to the direction of the grain. The cutter should comb the grain, not tear it up. When you cut against the grain, there is a greater chance of kickback.

Some woodworkers put a stationary fence on the table and feed the work between the fence and the cutter. In this technique, the stock is pinched between a moving cutter and a stationary object. This increases the chances of a kickback. This method of jointing should only be done with the aid of a power feeder (Illus. 6-27–6-32). The power feeder will minimize the chance of kickback and keep your hands away from the cutterhead. Also, use a dust-collection system to make this operation more efficient.

Illus. 6-28. Position a straight-edge fence the desired distance from the arc of the cutter. This will be the width of the stock after you have jointed it.

Illus. 6-29. Position the power feeder adjacent to the cutter and fence. The power feeder will feed the stock in through the opening between the fence and cutter. The wheels on the power feeder are in position so that they push the stock slightly against the fence and hold it down on the table.

Illus. 6-30. Turn the cutterhead on the same time you turn the power feeder on. Then position and feed the stock into the power feeder, which then feeds it through the cutter and pulls it out the outfeed side.

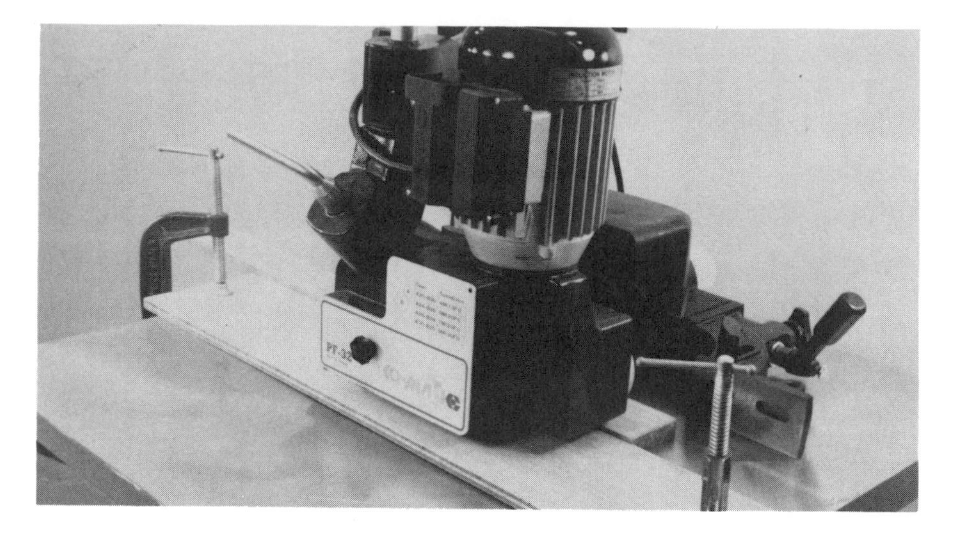

Illus. 6-31. Stock can be jointed rapidly with this power-feed setup. The speed of the power feeder is set so that a minimum amount of tearout occurs. This setup in most cases will give you a better cut than a jointer will, and keeps your hands away from the cutters.

Illus. 6-32. As you look through the barrier guard mounted to the spindle, you can see the amount of wood being taken with this power feed. This cut would be quite heavy if fed by hand, so you can see the benefits of the barrier. Also, note how difficult it would be to have contact with the cutter with this setup.

Cutting Rabbets

Rabbet cuts are L-shaped channels along the end or edge of a piece of stock (Illus. 6-33 and 6-34). A straight cutter is used to make rabbet cuts on the shaper. The width of the rabbet is controlled by the position of the fence relative to the cutter. The depth of the rabbet is controlled by the raising or lowering of the shaper spindle. This changes the position of the cutter relative to the table.

Illus. 6-33. A rabbet is an L-shaped channel along the edge of a piece of wood.

Begin the setup for a rabbet cut by mounting a straight cutter (Illus. 6-35) on the spindle. In

Illus. 6-34. This rabbet is along the end grain of a piece of wood. The L-shaped channel actually positions the drawer side relative to the drawer front. The drawer front would be the piece on your left, the drawer side the piece on your right.

most cases, the straight cutter should be as wide as the rabbet is deep, but in this case, that is not necessary. You can raise the cutter for a second or third cut to attain the desired depth.

After mounting the cutter, position the fences. The width of the rabbet determines how much cutter will extend beyond the fences (Illus. 6-36). Move the fences as close to the cutter as possible. It may be desirable to cut a notch or cope in the fences to accommodate the cutter. This will allow the fences to meet over the cutter and reduce the chance of the piece tipping into the opening between the fences. Lock the fences securely after adjusting them.

Illus. 6-35. Rabbeting requires a straight cutter, just as jointing does. A portion of the cutter is held below the table so that the desired rabbet is set.

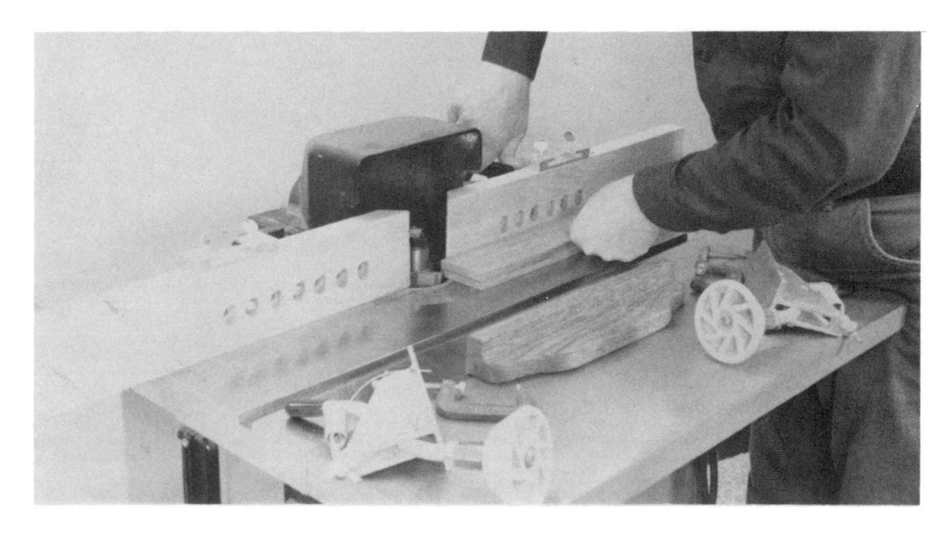

Illus. 6-36. The width of the rabbet determines how much cutter extends beyond the fence. Position the fence and lock it securely. It may be desirable to take more than one cut if it is a deep rabbet. Set a light cut first, and then move the fence a second time to the desired depth.

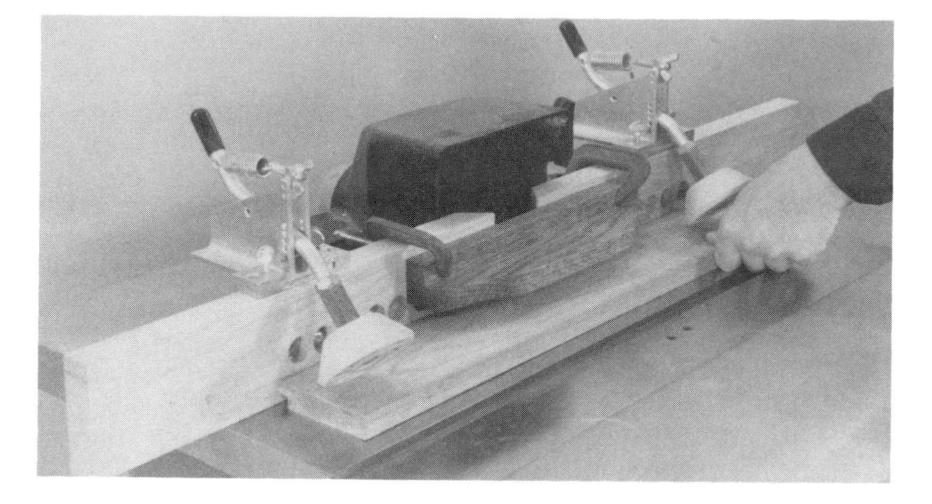

Illus. 6-37. This hand-fed rabbeting cut is being made using hold-downs and a barrier guard. Note how the operator stands to the side, clear of the kickback zone. Feed only as fast as the cutter will remove stock. This will minimize tearout.

Illus. 6-38. When using the power feeder, mount a special piece of stock against the fence to pick up the area where the rabbet is. This stock supports the workpiece uniformly on the outfeed fence. In many cases when using power feeders, you have to attach some kind of stock to the fence so that the bearing pressure is uniform on the outfeed side.

Illus. 6-39. The wood on this mitre gauge has been cut out to minimize tearout of the end grain.

Illus. 6-40. Stock has been clamped to this mitre gauge and positioned against the fence for end-grain rabbeting.

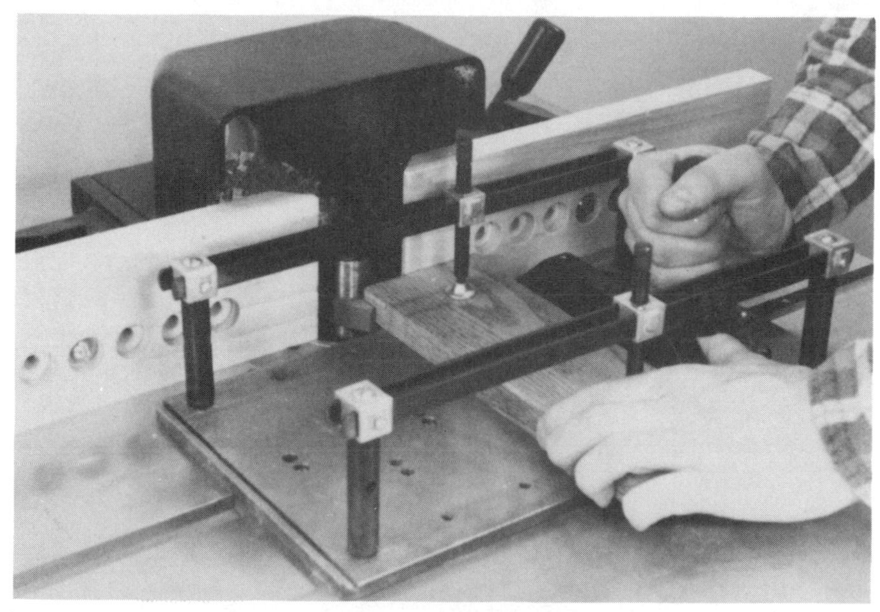

Illus. 6-41. A special mitring gauge is being used here to control stock while the end grain is being shaped. Note the clamping devices that hold the stock securely in position. It is important when rabbeting end grain that the fences be close together, close to the cutter.

Illus. 6-42. Two rabbets can be used side by side to make an edge- or end-grain joint. Some types of siding and flooring have this type of joint. When it is used in flooring, it is sometimes referred to as shiplap flooring. The rabbet is on alternate faces of each piece.

Cut rabbet using the same procedures explained earlier for end- and edge-grain shaping (Illus. 6-37–6-41). Make a practice cut on scrap stock. Make sure the setup is correct before shaping the work.

It may be necessary to take two light cuts to obtain the desired rabbet depth. This is done by elevating the spindle after the first cut. Also, note that rabbeting is done under the board. This allows the wood to act as a guard.

Cutting Edge and End Joints

Simple edge and end joints are cut in a manner similar to cutting rabbets. Some of these joints require two cutters, while others require only one cutter. Examined below are the setups for cutting different joints on the end or edge of the boards.

RABBET JOINTS

Rabbet joints are made by cutting rabbets on opposite edges or ends of the work. The rabbets are cut to one half the depth of the workpiece. For example, a board ¾ inch thick would have a rabbet ⅜ inch deep.

The width of the rabbet is equal to its depth. So, in the example just cited, the rabbet would be ⅜ × ⅜ inch.

Cut the rabbet on one edge from the best face of the workpiece. Cut the rabbet on the opposite edge from the back or poorer face. The ends are cut in the same fashion. Commercial siding and flooring that has edge rabbet joints are marketed as shiplap (Illus. 6-42).

Illus. 6-43. The V joint shown here is made with a V cutter and requires two separate setups: one for the male portion, one for the female portion.

Illus. 6-44. Position the V cutter at the centerline of the workpiece. Position the fences that the V cutter cuts to the intersection of the edge and face of the workpiece.

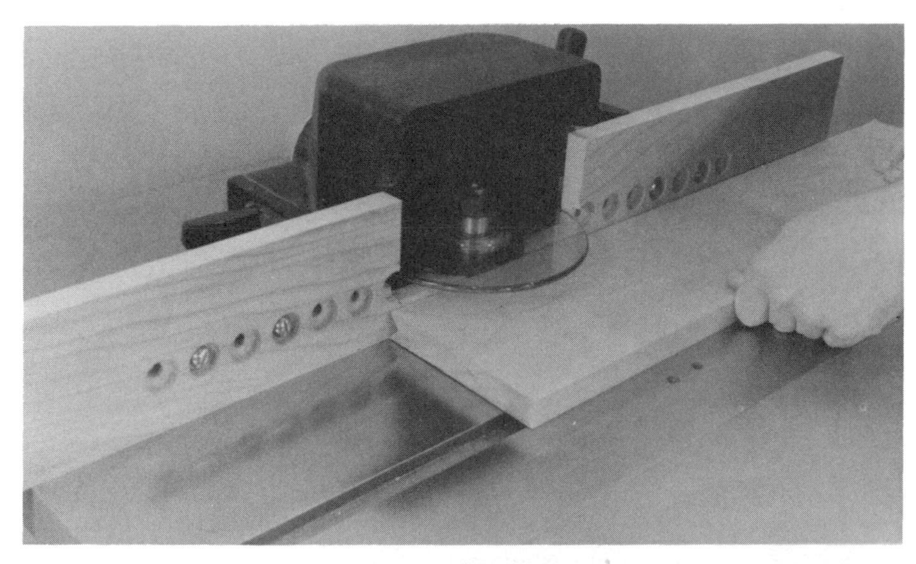

Illus. 6-45. Shape one edge of all pieces to this profile. Hold-downs and a push stick would be helpful in this setup, although there is a barrier guard to protect the worker from coming into contact with the cutter.

Illus. 6-46. Set up the mating cut by lowering the cutter and repositioning the fence. This setup removes a triangle of stock on the underside of the workpiece. This triangle extends from the centerline of the edge to the face of the work.

V JOINT OR U JOINT

A V joint requires one cutter and two setups (Illus. 6-43). Mount a V cutter at least as thick as the work on the shaper. For the first setup, position the center of the V cutter at the centerline of the workpiece (Illus. 6-44). Position the fences so that the V cutter cuts to the intersection of the edge and face. Shape one edge of all boards to this profile (Illus. 6-45).

After shaping one edge, lower the cutter and reposition the fence. Set up the mating cut. The cutter cuts a triangle off the underside of the board (Illus. 6-46). This triangle extends from the centerline of the edge back through the face of the work.

Shape the work on the opposite edge. Two cuts will be required. Cut the work on one face; then turn it over and cut the other face (Illus. 6-47). The edges should join perfectly (Illus. 6-48). This type of joint is frequently used for edge-banding plywood (Illus. 6-49–6-51).

A U joint (Illus. 6-52) is essentially the same as a V joint. This joint requires two mating cutters that are as wide or slightly wider than the stock is thick. Cut one edge of all the pieces with a concave cutter (Illus. 6-53–6-55). Cut the other edge with a convex cutter. The setup for this cut should be oriented off the centerline of the work.

When cutting a U shaped joint, you may have to offset the fences on the convex edge of the

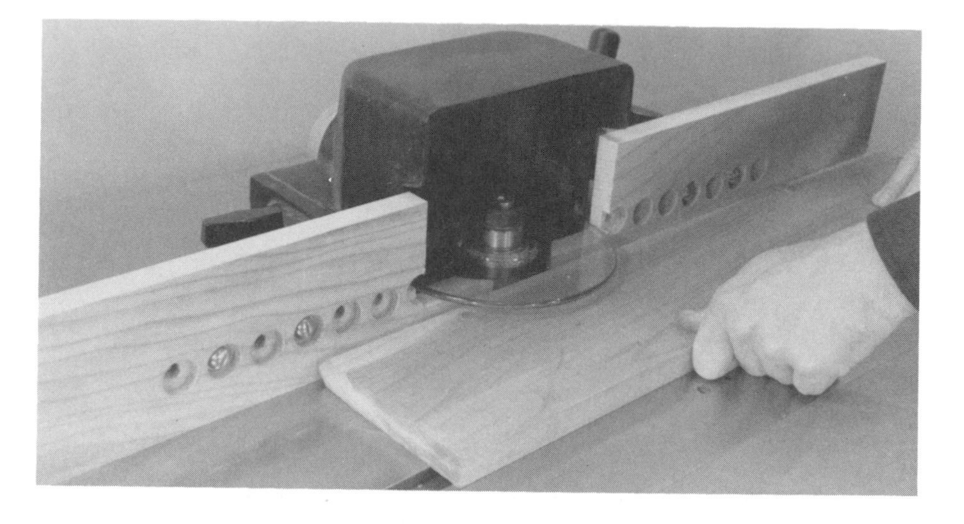

Illus. 6-47. Turn the stock over, and make the same cut on the opposite face of the workpiece.

Illus. 6-48. Make a test cut on a piece of scrap to be sure that the joint is perfect before shaping all the stock. To ensure the best results, work carefully and check your layouts.

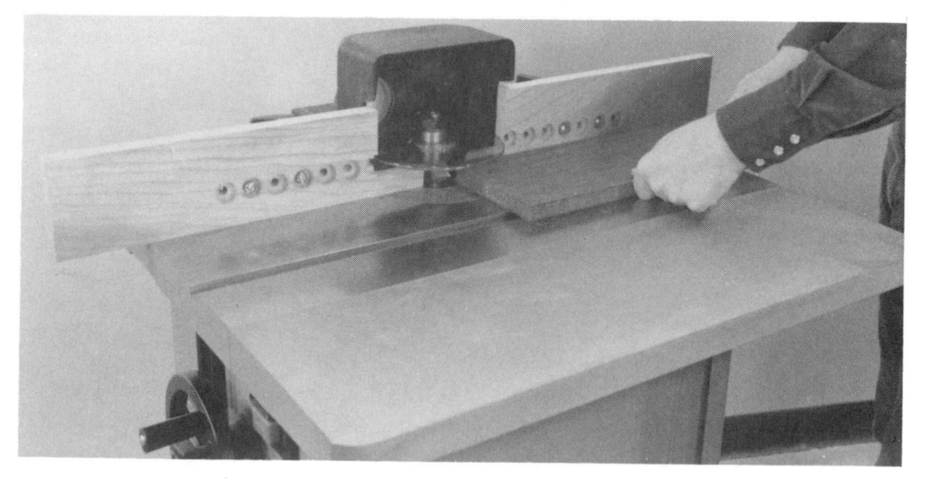

Illus. 6-49. You can edge-band plywood by making a V cut in the edge or end of the plywood.

Illus. 6-50. A V cut is being made on the edge of this piece of plywood shelf stock. This cut will make it possible to edge-band the stock.

Illus. 6-51. A square piece of edge-banding has been glued into the V cut made on the edge of this piece of plywood. This technique will work on all types of plywood. This square stock can later be shaped into some decorative profile, if so desired.

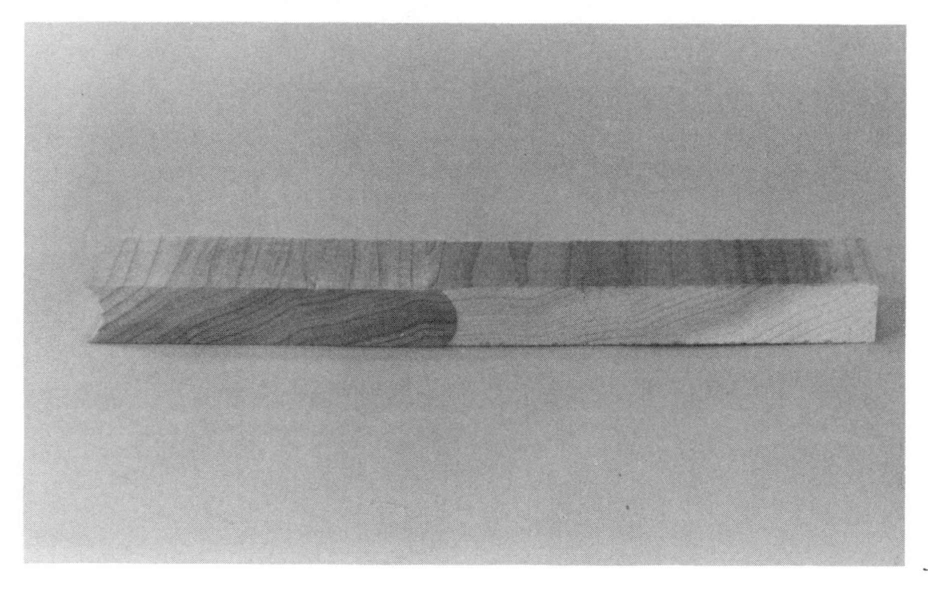

Illus. 6-52. A U joint is similar to a V joint, except that a U joint will pivot slightly when clamped. This means that the piece can be turned, which is desirable when you are doing tambour work and for boat hulls.

Illus. 6-53. First, set up a concave cutter on the shaper spindle. Center the cutter on the workpiece.

Illus. 6-54. Align the fences with the cutter, and clamp a barrier guard to the fences. Now make the concave cut.

Illus. 6-55. When making the concave cut, use hold-downs to hold the stock down on the table and in against the fence.

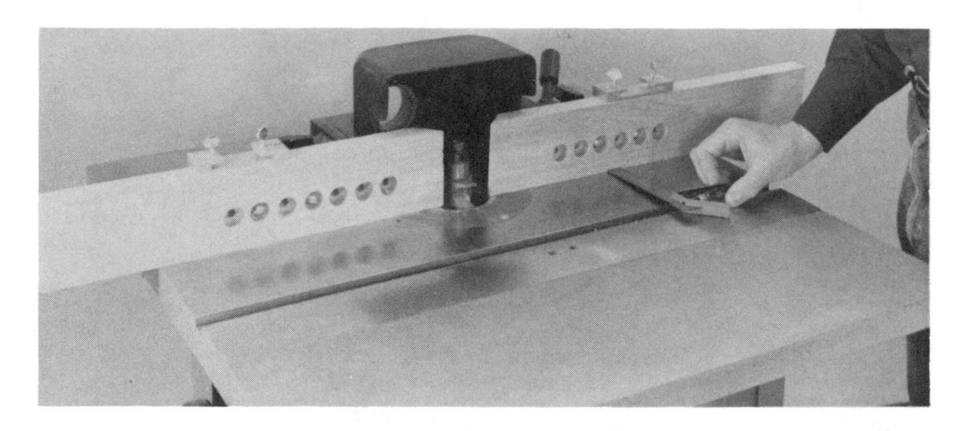

Illus. 6-56. If you are going to shape the end grain for this joint, then you must align both fences with the mitre slot before beginning. First, align the infeed fence with the mitre slot as shown here.

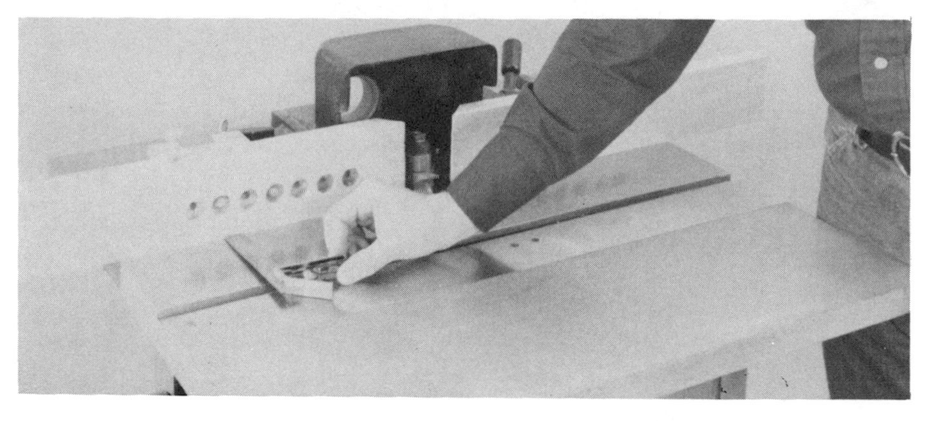

Illus. 6-57. Next, align the outfeed fence with the mitre slot.

Illus. 6-58. Align the outfeed fence with the full depth of the cutter.

Illus. 6-59. Then offset the infeed fence so that you can remove the entire edge of the board.

Illus. 6-60. A piece of stock can help you lay out and set up the position of the cutter and the infeed fence.

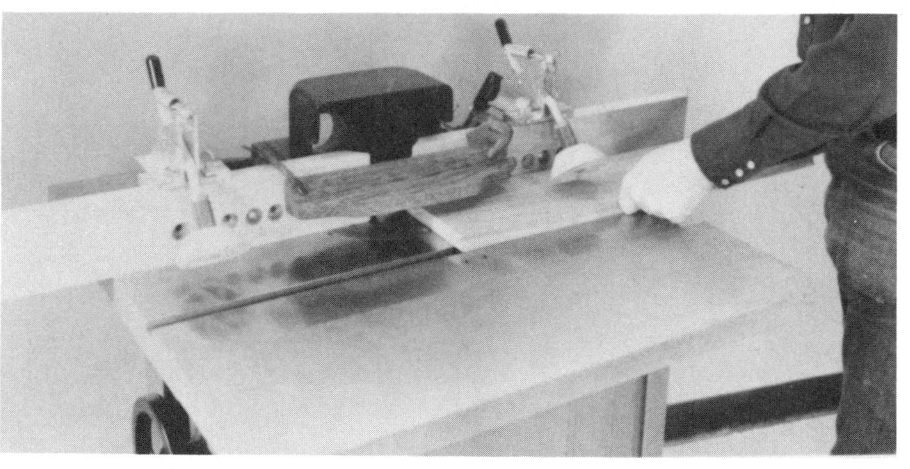

Illus. 6-61. The convex cut is similar to the concave cut. It is just an edge-joint cut in which you use the fences for supporting control. Barrier guard hold-downs can be useful here.

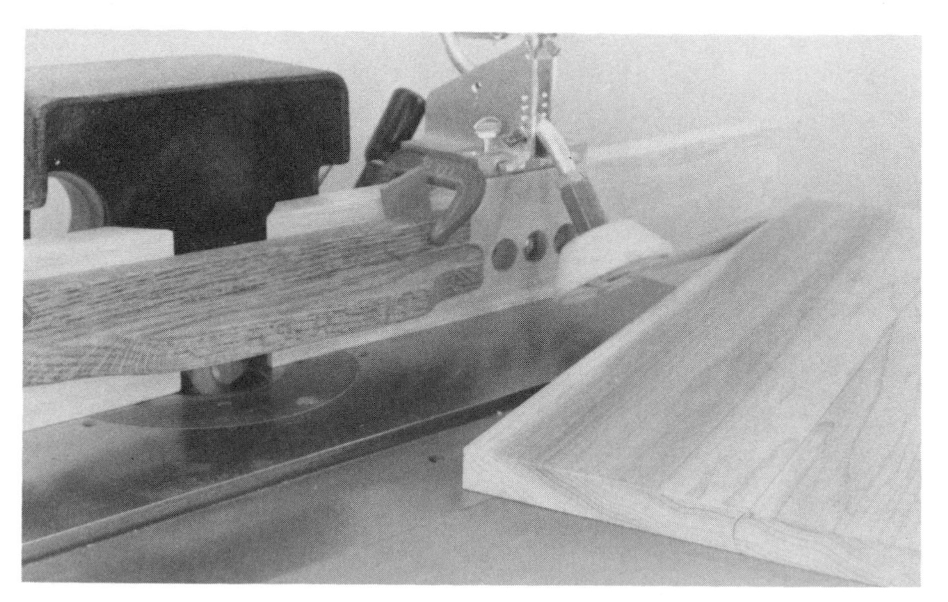

Illus. 6-62. The joint fits together very well. Note how tightly the two pieces join.

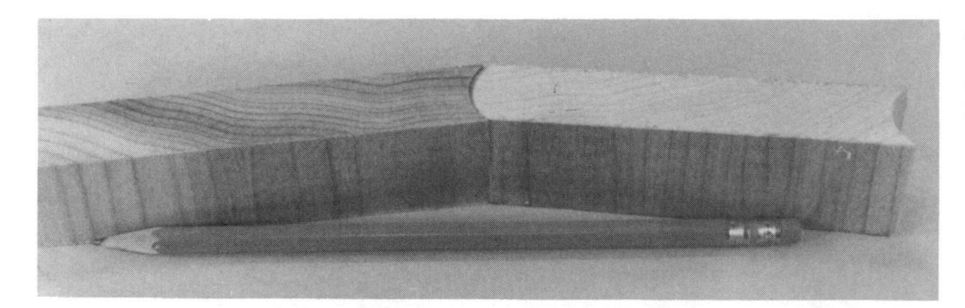

Illus. 6-63. When a U joint is used, the stock can be bent or pivoted around an irregular object.

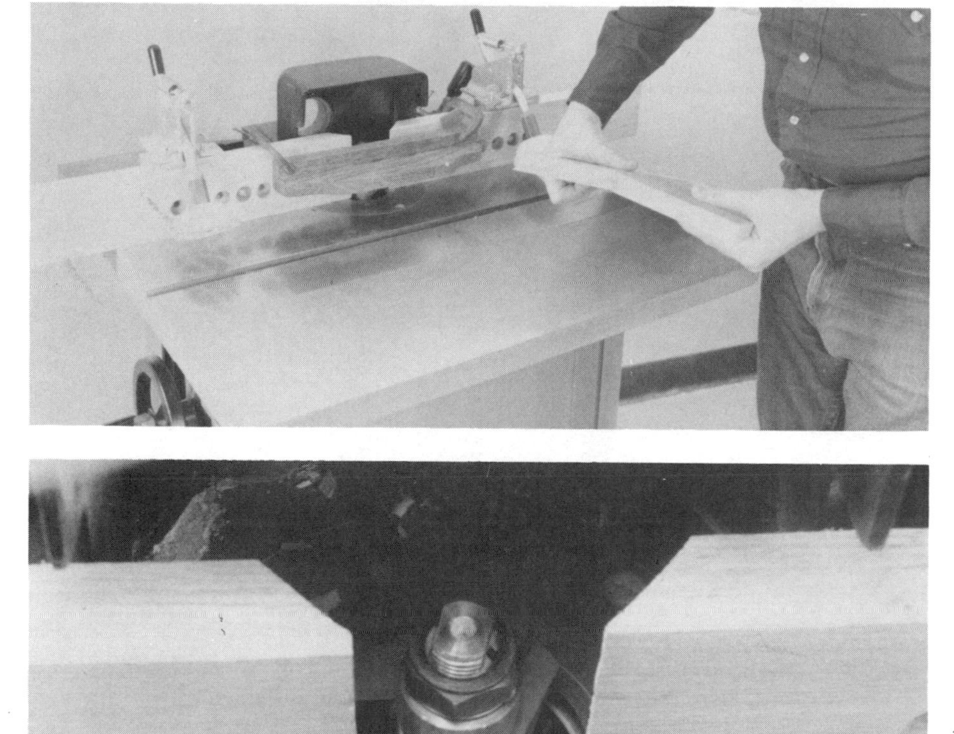

Illus. 6-64. When the pieces are bent like this, the total effect would be the same as that of a rolltop desk or tambour type of door.

Illus. 6-65. Plywood that has been edge-banded with square stock can be rounded with a convex cutter. Note how the fences have to be offset for this operation.

piece (Illus. 6-56–6-62). The joint fits snugly, and the pieces form an arc while still remaining in close contact. This joint is frequently used for tambour parts on rolltop desks and other products (Illus. 6-63 and 6-64).

With the same setup, you can round plywood that has been edge-banded with square stock (Illus. 6-65–6-68). This operation requires offset fences. Use a barrier guard over the cutter to make the operation safer. The cutter removes the sharp point and improves the appearance of the shelf edge. If the square plywood had been clamped in position, the clamps may have damaged the thin point. This shaping operation removes any damage caused by clamps.

Another shaping approach is shown in Illus. 6-70. The solid stock shown was edge-glued to the plywood and then shaped with a smaller cutter. This left a bead. The thicker edge band makes the shelf look thicker. This is one way to make shelves of equal proportions.

Larger beads or radii can be shaped with the appropriate cutters (Illus. 6-71). You must align these cutters correctly before beginning (Illus.

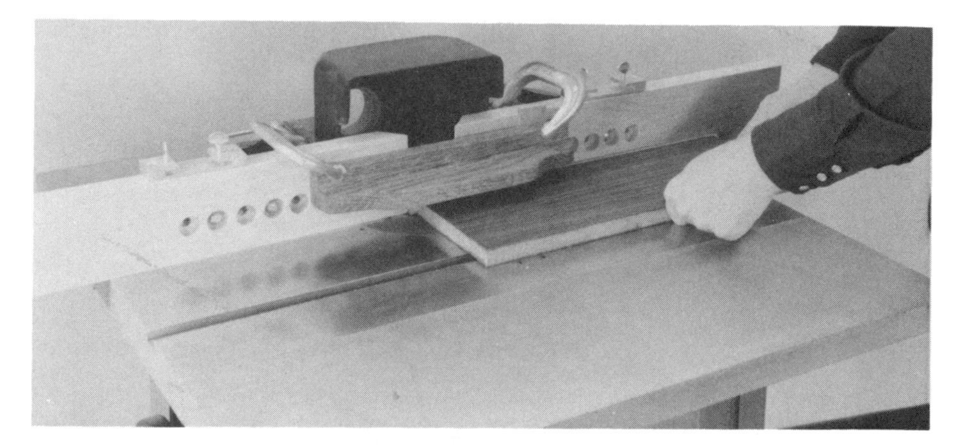

Illus. 6-66. This plywood shelving stock is being shaped to remove the square edge from the edge-banding.

Illus. 6-67. A barrier guard has been positioned over the stock to prevent contact with the cutter.

Illus. 6-68. The radius formed on the edge-banding is far more decorative than the square edge. Rounding the plywood eliminates any blemishes that might have been caused by clamping.

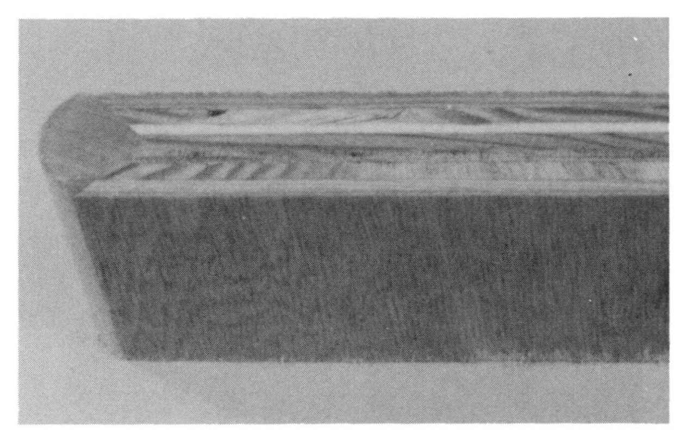

Illus. 6-69. You can use veneer-core, fibre-core, or lumber-core plywood for the edge-banding.

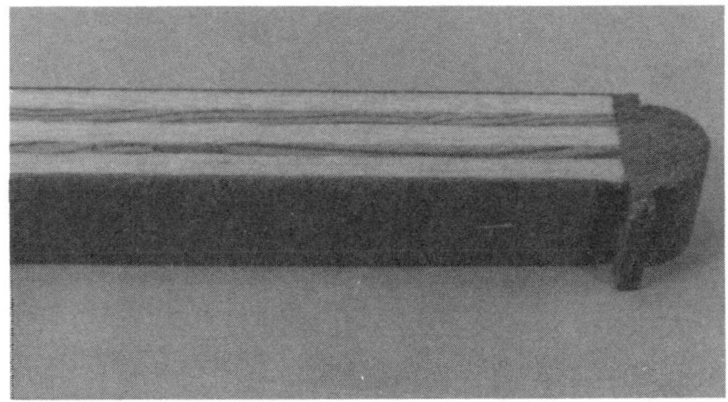

Illus. 6-70. A thick edge-band can simply be face-glued to the plywood and shaped later. This makes the shelf look thicker than it actually is, which is an effective technique for some shelving jobs.

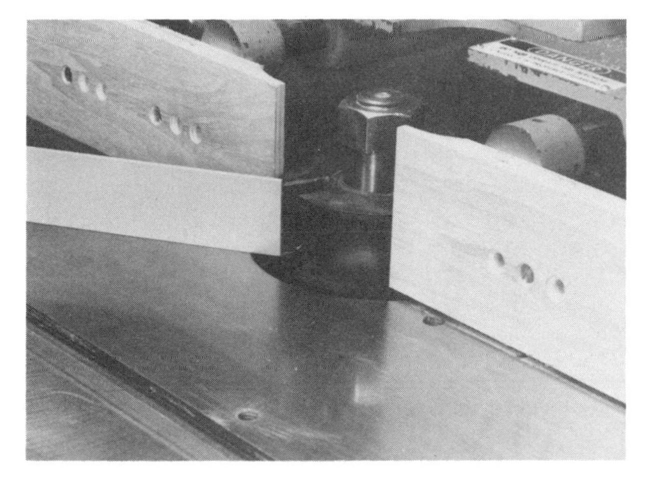

Illus. 6-71. A large concave cutter can be used for bullnosing operations—that is, when shaping large beads. Adjust the lower edge of the cutter so that it's even with the table.

Illus. 6-72. Position the concave cutter so that it is as deep as the depth of the outfeed fence.

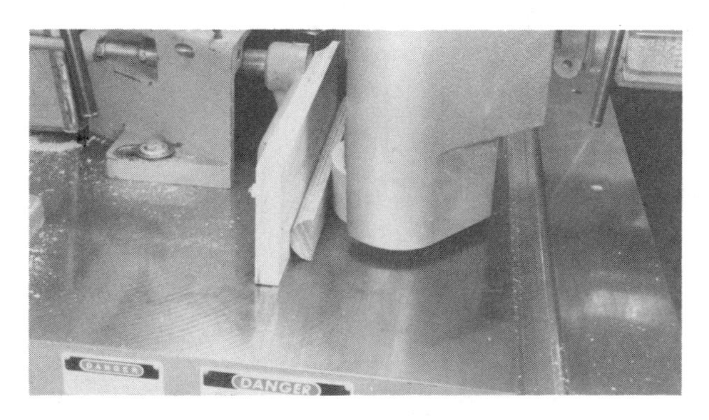

Illus. 6-73. A bullnosed cut like this requires a power feeder. Stock this size cannot be handled safely in a hand-feeding operation.

Illus. 6-74. Tongue-and-groove joints are made using two complementary cutters. The pieces fit together to make panelling or flooring.

6-72). These heavy cuts are best made using a power feeder (Illus. 6-73). Power feeders keep your hands clear of the cutterhead, reduce the chance of a kickback, and feed the stock uniformly.

TONGUE-AND-GROOVE JOINTS

Tongue-and-groove joints (Illus. 6-74) require two complementary cutters. One edge or end of the board forms a tongue; the other edge forms a groove. The centerline of the workpiece is generally used for all setups (Illus. 6-75). In the case of flooring, the tongue and groove can be located below the centerline. This gives the upper side of the groove more strength and resistance to splitting. It also allows for sanding of the floors after prolonged wear.

Illus. 6-75. Set up the tongue cutter first. It usually is oriented off the centerline. It sometimes may be below the line, as when flooring is being cut.

When cutting the tongue portion of the joint, you may have to offset the fences to remove the stock (Illus. 6-76 and 6-77). Remember to keep the fences close to the cutter.

Twice as much stock is removed when the tongue is cut, so you may have to make two light cuts. The groove can be made with a single cut.

Use the tongue part of the joint to set up the groove cutter (Illus. 6-78). Be sure to keep the good (or bad) face up during all shaping. Use hold-downs to control the stock (Illus. 6-79 and 6-80). This will ensure that the parts form a true plane when they are joined together.

You can also cut tongue-and-groove joints on end grain. If making this joint on long pieces, make sure that the pieces are well supported during the cut. Cut the end-grain joints first.

Some tongue-and-groove cutters are straight. Others are wedge cutters. Those shown in Illus. 6-80 are wedge cutters. Their edges are tapered, and, as the pieces they cut are put together, there is a wedging action. The wedging action can cause splitting if the fit is not correct. This

Illus. 6-76. To completely remove stock when making the tongue, offset the infeed table.

can be a problem if you are installing the pieces as flooring or panelling.

The straight tongue-and-groove joint shown in Illus. 6-74 has a bevel cut on it. This type of joint is usually cut on wall panelling. The bevel gives the panelling more line. Another form of panelling is presented later.

GLUE JOINTS

A glue joint is similar to a tongue-and-groove joint, but uses only one cutter. It is oriented off the centerline of the workpiece. One edge is cut with the good face up. The opposite edge is cut with the good face down. The two edges then mate to form a true plane.

Glue joints are used to join parts edge to edge. They are also used to fasten edge bands to plywood.

When shaping glue joints, make sure that the stock is held down on the table and in against the fence. If the parts wander away from the fence or table, the joints will not fit correctly. Use the necessary hold-downs to control the

Illus. 6-77. Use hold-downs and a barrier guard to make the operation safer when shaping the tongue.

Illus. 6-78. Use the tongue cut to set up the groove cutter. Line up the point of the tongue with the centerline of the groove cutter. This matchup will allow stock to be shaped correctly.

Illus. 6-79. Cut the groove on the opposite edge of the workpiece. Make sure that the same face is against the shaper table when making this cut.

Illus. 6-80. When the shaping is completed, the stock should wedge together and fit securely. You can also shape end grain using a mitre gauge.

Illus. 6-81. When setting up a glue-joint cutter, make sure that you have the appropriate accessories on hand before you begin.

Illus. 6-82. Make sure that the glue-joint cutter is turning in the proper direction, that the appropriate table insert is in position, and that the spindle nut is locked securely. If two spindle nuts are used, be sure to put them both on.

Illus. 6-83. Align the outfeed fence with the deepest portion of the cut on the glue-joint cutter.

Illus. 6-84. Slightly offset the infeed fence so that you can joint the stock's entire edge.

Illus. 6-85. Move the fences as close to the cutter as possible to reduce the mechanical advantage the cutter has over the workpiece.

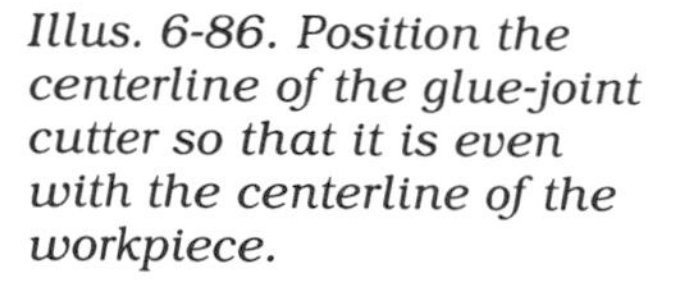

Illus. 6-86. Position the centerline of the glue-joint cutter so that it is even with the centerline of the workpiece.

Illus. 6-87. Since these are mating pieces, make sure that the good face of one workpiece is down when you shape the piece.

Illus. 6-88. Shape the other mating piece with its good face up. Note how the hold-down is positioning the stock against the fence and acting as a barrier between the operator's hand and the cutter.

Illus. 6-89. Test the setup on some scrap stock before actually shaping the work. The stock should align evenly on the top and bottom. If not, check the stock for uniform thickness before you re-adjust the height of the cutter.

Illus. 6-90. Some glue joints actually have straight sides. The cutters used to make these joints also work off the centerline of the stock.

work. The sequence for making a glue joint is presented in Illus. 6-81–6-89.

Some glue joints have straight sides (Illus. 6-90). In this case, also work the cutters off the centerline of the work (Illus. 6-91). If you are not going to joint the stock, do not offset the fences. Recess the infeed fence slightly (Illus. 6-92). This way, the cutter will joint the edge before the joint is cut.

To make mating joints, make sure that the good face of one piece is up, and the good face of its mate is down (Illus. 6-93). Since the parts are mirror images of each other, the fit should be perfect (Illus. 6-94).

If the faces of the two pieces do not line up, check the pieces to be sure they are equal in thickness. If they are equal, then the cutter height must be adjusted by one half the offset of the stock faces. Once the setup is adjusted, test it on a piece of scrap (Illus. 6-95).

Illus. 6-91. This glue-joint cutter is actually a router bit in a special collet on the shaper. A plywood base has been mounted on the table to raise the stock to the appropriate spot on the cutter. The stock is positioned relative to the centerline of the workpiece.

Illus. 6-92. Line up the fences so that no stock is actually removed from the edge. If you want to joint the stock, you can offset the infeed fence slightly.

Illus. 6-93. Use a barrier guard over the bit to eliminate contact with it. Hold-downs would improve the safety of this operation. Note how the operator stands to the side of the cut.

Illus. 6-94. The mating edge-joint pieces fit snugly if the cut has been centered. If your workpiece does not line up correctly, check to be sure the stock is of a uniform thickness before adjusting the height of the bit.

Illus. 6-95. Always test the setup on scrap stock before actually shaping the stock. This will minimize waste.

FINGER JOINTS

Finger joints are another type of edge joint (Illus. 6-96–6-100). The mating pieces for this joint should be shaped with opposite faces of the workpiece on the table.

Finger joints work well on end or edge grain. They also work on certain types of plywood.

Keep the fences close to the cutter when using the finger joint cutter. Use a power feeder on thick stock for greater control. Work off the centerline of your workpiece. Test-cut several pieces to get the best setup.

Illus. 6-96. Finger joints are another type of edge joint. There is a large amount of surface on this joint for glue, which makes it very strong.

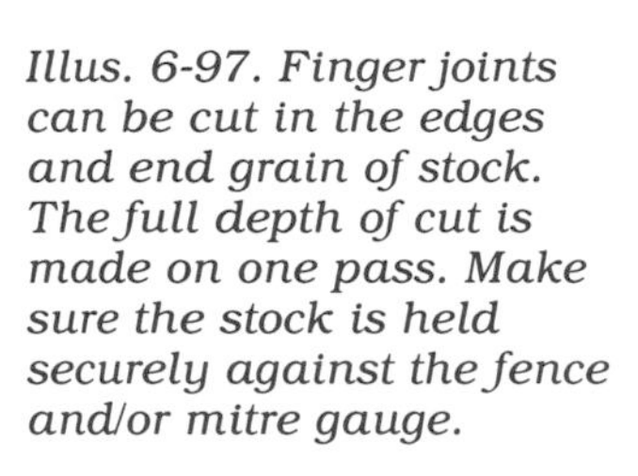

Illus. 6-97. Finger joints can be cut in the edges and end grain of stock. The full depth of cut is made on one pass. Make sure the stock is held securely against the fence and/or mitre gauge.

Illus. 6-98. Finger joints can be cut in edge or end grain, as well as in plywood.

Illus. 6-99. The cutter height must be adjusted to cut the mating joint. Align the tips of the cutter with the tips of the mating cut.

Illus. 6-100. This mitred finger joint makes a decorative frame that will not come apart.

Making Wall Panelling

You can make wall panelling by using a series of straight cutters and spacers (Illus. 6-101). The cutters should all be equal in diameter, but vary in width of cut. The spacers should be equal in thickness and diameter.

Adjust the fences for the depth of cut (Illus. 6-102). Keep the fences close to the cutters. Set up the power feeder (Illus. 6-103) so that it will feed against the fence. *Note:* It is important that the fence be locked securely to the table.

After all the faces have been shaped, rabbet one edge of all the pieces (Illus. 6-104). Adjust the height of the spindle accordingly. Set up the power feeder so that it will feed against the table (Illus. 6-105), and shape one edge on all the panelling.

Adjust the fit of the parts so that the rabbet hides any gap or nails when the pieces are installed (Illus. 6-106). You can install this panelling horizontally, vertically, or at an angle. For a different effect, use wood of various species.

The best feature of this shaping method is that both cuts can be made with a single spindle height adjustment. If you use a power feeder, you will be able to shape the pieces quickly, and all parts will be uniform. The shaper is probably the most productive of the woodworking machines.

Improving Shaped Joinery

As you work with the shaper and various cutters, you will be pleased by the quality of the joints you have made. However, sometimes you will find that shaped joints do not always fit together as well as you had hoped. So, pay close attention to the following guidelines, and you

Illus. 6-101. This series of straight cutters and spacers has been mounted on the spindle to make a special kind of panelling. The cutters are all equal in diameter, but vary in width of cut.

Illus. 6-102. Position the outfeed fence for the desired depth of cut in the face of the panelling.

will be able to set your shaper up for quality joinery on the very first attempt:

1. Lay out the stock carefully.
2. Set up the shaper carefully and lock all the settings securely.
3. Test all setups in scrap stock before cutting your work. Keep your setup simple. It will reduce setup time next time you use the same cutters.
4. Look over your stock. Stack it in the way you wish to shape it. This will minimize the chance that you will make an error.
5. Control the stock. Support long pieces and use hold-downs to keep the stock against the table, fence, or mitre gauge.
6. Check the parts periodically. This will help you identify any setup problems early.

Illus. 6-103. Set up the power feeder so that it shapes the face of the panelling. Note how the power feeder has been positioned parallel to the panelling and the fence.

Illus. 6-104. After shaping all the faces, re-adjust the cutterhead so that you can cut a rabbet on one edge of all of the panelling pieces. Note the setup: Both fences are in alignment because the entire edge of the piece is not being removed.

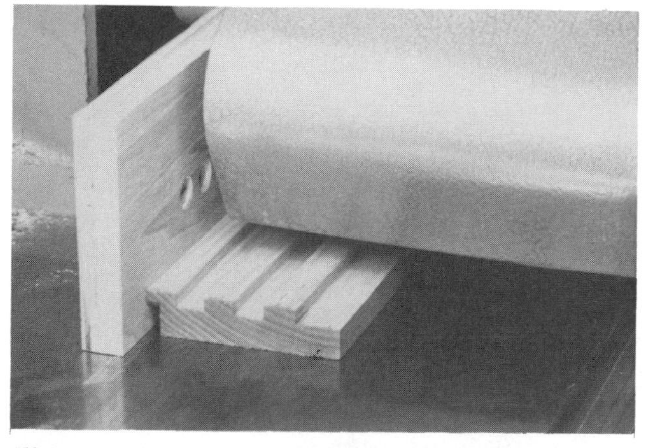

Illus. 6-105. Note how the power feeder can be used to shape this edge as well. The rabbet is cut without the cutters on the head, and the power feeder acts as a barrier between the operator and the cutter.

Illus. 6-106. The rabbet that is cut on one edge of the panelling actually hides the nails after the panelling has been installed. This panelling can be installed vertically, horizontally, or diagonally.

7
INTERMEDIATE AND ADVANCED SHAPER OPERATIONS

Before attempting any operation in this chapter, make sure that you are familiar with the operations presented in earlier chapters. This chapter requires experience and knowledge beyond beginner status. Master the techniques explored in Chapter 6 before attempting operations in this chapter.

Freehand Shaping

Freehand shaping consists of moving straight or irregular shaped pieces of wood through the shaper by hand. The stock is controlled with rub collars. Hold-downs and starting pins support the work and make the operation safe. These devices reduce the chance of kickback.

The stock can also be controlled with jigs, fixtures and vacuum hold-downs. Jigs, fixtures and vacuum hold-downs ride against the rub collar and control the depth of cut.

Freehand shaping provides many opportunities to enhance a simplified design and create more sophisticated cabinetry and furniture. Freehand shaping should only be attempted by experienced woodworkers. For best results, take light cuts and work carefully. Be sure to guard the cutter whenever possible.

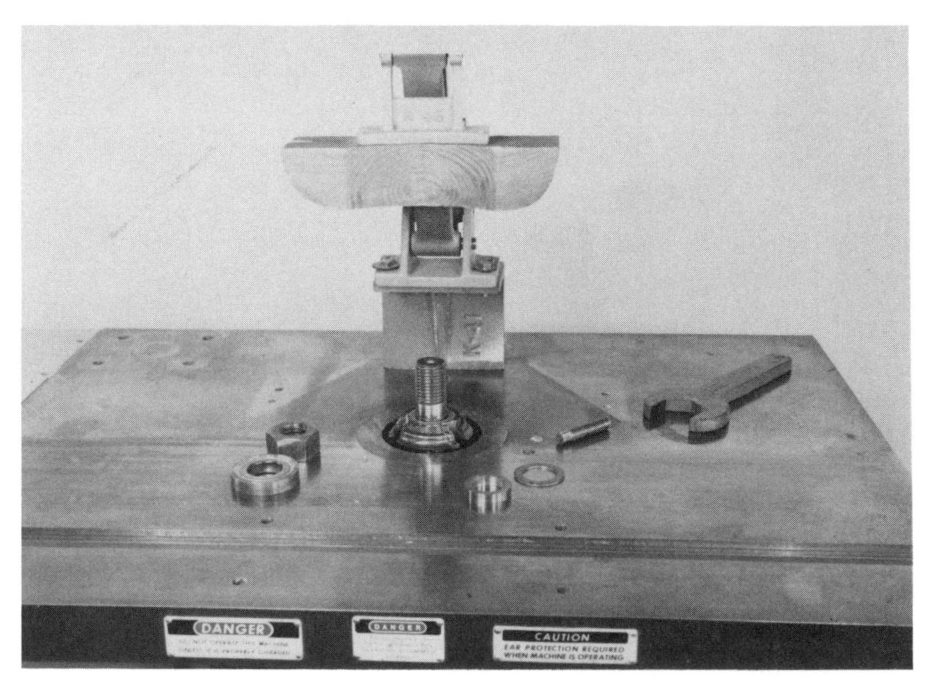

Illus. 7-1. This shaper is being set up to shape small irregular mouldings.

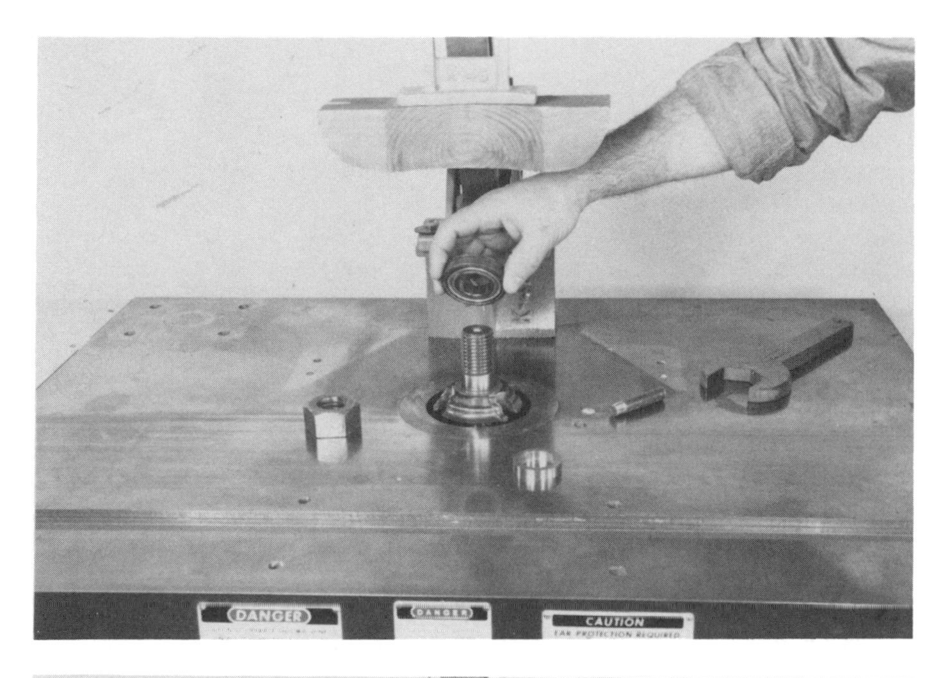

Illus. 7-2. Position a shim or spacer between the cutter and the ball-bearing rub collar.

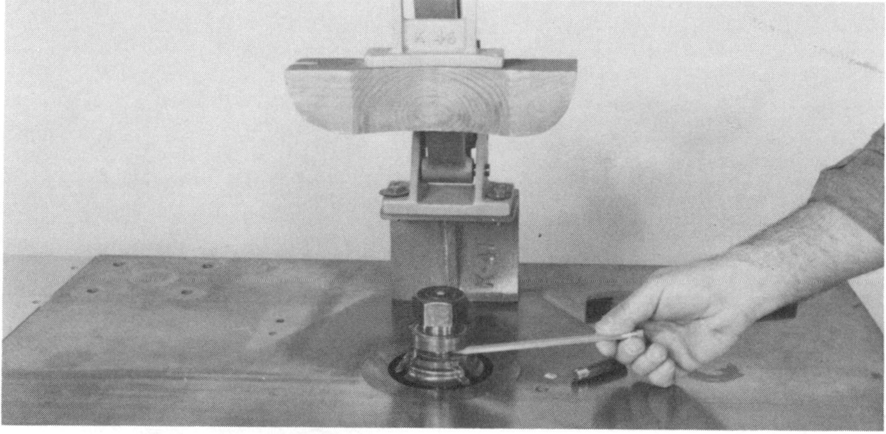

Illus. 7-3. The rub collar will control the depth of cut and the path of the workpiece.

Illus. 7-4. Position the starting pin on the thrust side of the cutter.

Illus. 7-5. Adjust the height of the cutter so that it makes a light cut. This reduces the chance of a kickback.

Illus. 7-1–7-8 show and describe how to freehand shape small mouldings. Illus 7-9–7-22 show how to freehand shape long window mouldings. Illus 7-23–7-26 show how to freehand shape a curved handrail.

The safety procedures for freehand shaping are consistent with the general safety procedures outlined in Chapter 4. You should also consider the direction of feed (whether the cutters will be used in a reverse or forward rotation) and have an understanding of the relationship between rpm (revolutions per minute) and the diameter of the cutter.

Illus. 7-6. The hold-down also acts as a barrier guard in this setup. Butt the workpiece against the starting pin and move it into the rotating cutter.

SETTING UP THE RUB COLLAR

Rub collars act as a guide or pivot for the work to ride on. They can be located either above or below the cutter. A rub collar controls the depth of cut because it has the same diameter as the smallest diameter on the cutter profile.

Illus. 7-7. Once the work has engaged with the rub collar, pull it away from the starting pin.

Illus. 7-8. Make the second cut after raising the spindle. Two light cuts are better than one heavy cut.

Illus. 7-9. Lift the ball bearing and the metal sleeve collar off the arbor with two wooden blocks and the elevating mechanism.

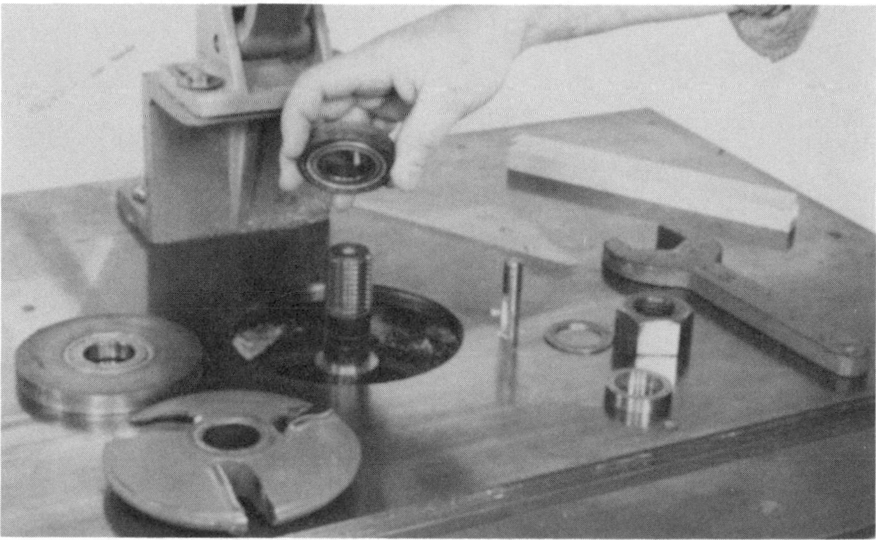

Illus. 7-10. Mount a smaller rub collar onto the arbor. This setup will produce curved window mouldings. Note: The rub collar will be under the workpiece for this setup.

Illus. 7-11. Place a shim between the collar and the cutter. This ensures that the bearing will turn freely.

Illus. 7-12. This cutter can be mounted on the arbor in two different ways. Each side has a different cutting profile. Note that this cutter is a chip-limiting cutter. This reduces the chance of kickback when you are freehand-shaping.

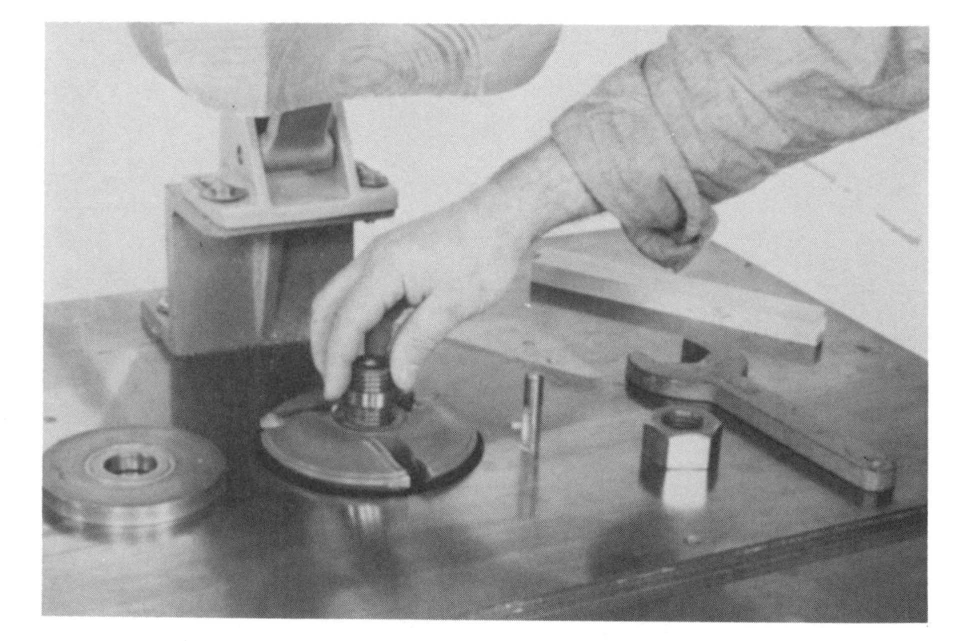

Illus. 7-13. Position a spacer over the cutter, and then secure the arbor nut.

Illus. 7-14. The starting pin helps line the workpiece up with the cutter and rub collar.

Illus. 7-15. Position the hold-down and begin shaping.

Illus. 7-16. This workpiece is quite long. The hold-down holds the work while the operator moves to the other side. Note the take-off table that supports the workpiece.

Illus. 7-17. Pull the work across the shaper cutter. The stock is being controlled with the rub collar and the hold-down.

Illus. 7-18. You must support long stock on both of its ends to obtain quality work. Otherwise, if the end of the workpiece drops down, the cutter would dig into the work and possibly cause a kickback.

Illus. 7-19. Remove the cutter and mount a new depth collar onto the arbor.

Illus. 7-20. The depth collar will ensure that only the outer edge of the cutter does the cutting.

Illus. 7-21. Adjust the height of the cutter for a second cut.

Illus. 7-22. The completed work. It is made with two cuts with the same cutter and different depth or rub collars.

Illus. 7-23. In this curved handrail setup, the rub collar will ride against a piece of plywood that has been nailed to the bottom of the workpiece.

Illus. 7-24. The rub or depth collar is so large that the full profile is not cut. Light cuts work best when you are freehand shaping.

Illus. 7-25. Replace the large rub or depth collar for the second setup.

Illus. 7-26. Cut the completed profile. Note that a hold-down has to be used.

Rub collars can be purchased with cutter sets or made with ball bearings (the inside diameter of the collar will be the same as the diameter of the spindle). Press steel or plastic sleeves onto the bearing to enlarge the outside diameter of the collar, if required, for different size cutters. The height of the collar can vary, depending upon the length of the spindle used. Both the cutter and the rub collar must be below the threaded portion of the spindle. This will keep the spindle nut from working loose, and eliminate any vibration.

Wood will burn or scar during the shaping operation if the rub collar does not spin freely. Place a shim between the collar and the cutter. This will provide clearance between the collar and cutter, and allow the rub collar to spin freely. A steel washer makes a suitable shim if its inside diameter is the same as the spindle diameter.

Make sure that the bearings used are sealed. Wood chips or dust cannot enter a sealed bearing. Wood chips and dust can reduce the life of the bearing.

Extended use of the bearing or collar may cause overheating, and prevent the collar from spinning freely. Check the bearings periodically to make sure that they are not worn. Replace the bearing at the first sign of wear. It is better to invest in a new bearing than to damage the work or cause an accident.

Rub collars can also be attached to knife holders. Chapter 3 explores this method of attaching rub collars.

When selecting bearings for rub collars, check to see if they rotate clockwise or counterclockwise. Some bearings are made for one direction only. Reversing the direction of the bearing may result in burning the workpiece, because the bearing cannot spin freely.

Carefully place the rub collar onto the spindle. Do *not* hammer them onto the spindle. Instead, always press them using the spacer rings on top of the bearing. The collars will slide down the spindle as you tighten the spindle nuts.

To remove a collar, first remove the spindle top nut and then the cutter. Raise the spindle until the collar has approximately ¾ inch clearance between the table surface and the bottom of the collar. Then place two hardwood squared sticks under the collar (Illus. 7-9) and lower the spindle, raising the collar to the threaded end of the spindle, where it can be removed by hand.

Rub collars are an important part in freehand shaping, and must be protected in the same manner as the cutters themselves. Well-maintained cutters and rub collars make for a better finished project.

For some profiles, two light cuts will be better than one heavy cut. In this case, use two rub

Illus. 7-27. Vacuum hold-downs can be useful when you are freehand-shaping small parts on one or both sides.

Illus. 7-28. A vacuum hold-down is made of nonporous material with a gasket around its perimeter.

collars of different diameters. Make the first cut using the collar with the larger diameter. Make the second cut using the rub collar with the smaller diameter. This rub collar moves the work closer to the cutter, which means you can make a second light cut. You can also make a second light cut by raising the spindle in some operations.

While the technique of making two light cuts requires an additional setup and handling of the work, it makes the operation safer and reduces the chance of kickback. Light cuts also increase feed speed, which reduces the chance that the stock will burn. *A note of caution:* The rub-collar bearing should engage with at least ¼ inch of stock thickness. With less of the rub-collar bearing engaged with the work, the work could slip off the bearing into the cutterhead. This could damage the work or cause a kickback.

USING THE STARTING PIN

Starting pins are threaded or tapered steel posts that are attached to the table surface. They are used as pivoting points for freehand shaping. The starting pin minimizes the chance of kickback. Kickbacks can occur when the shaper cutter grabs the end grain when the work is not supported.

Since work is fed into the rotation of the cutter, position the starting pin as close as possible to the cutter on the infeed side. Butt the work to the starting pin, and carefully feed the work into the rub collar. Start your cut on edge grain, not end grain. This will reduce the chance of kickback. Once the work is engaged with the collar, you should have no further contact with the infeed starting pin.

When you rely on the starting pin, the work sometimes leaves the bearing. If you push the work back to the bearing, there will be a kickback. If the work has left the bearing, pull the work completely away from the cutter and begin again, using the starting pin.

When working with small irregular workpieces, use two starting pins. One pin starts the job, and the other pin finishes the job. The height and diameter of the starting pins may vary according to the workpiece being shaped.

If you have split fences, you can remove the outfeed fence and adjust the infeed fence so that it can be used as a starting pin. By positioning the fence close to the cutter, you have added protection against kickback and tearout. In some cases, you can clamp a stop block to the fence. This helps position the stock when you begin to shape it.

FREEHAND SHAPING WITH A VACUUM HOLD-DOWN

When freehand shaping small parts, you'll discover that vacuum hold-downs can be helpful

(Illus. 7-27). Vacuum hold-downs are also useful when the workpiece needs to be shaped on both sides, or when the job requires quick removal and changing for production setups.

Vacuum hold-downs are shop-made devices that are similar to suction cups (Illus. 7-28 and 7-29). The shape of the clamp itself is made to conform to the size of the workpiece. The clamp board should be made of high-density particleboard, Baltic birch plywood, or other suitable nonporous material. Non-porous materials are used because air cannot pass freely through them. This ensures that the vacuum is maintained at all times.

Once you have cut the clamp board to size, route a ¼-inch groove around its bottom perimeter. Make the groove ⅛ inch deep by ¼ inch wide, the same size as standard ¼ inch by ¼ inch foam door seal. Place dense foam into the routed groove. Be careful not to stretch the material. Attach the adhesive side of the foam to the wood. Make sure that the cut ends fit together well. This prevents leaks and increases the clamping power of the vacuum hold-down.

Illus. 7-29. A vacuum hold-down acts like a large suction cup. It holds securely to the workpiece. The hose is connected to a vacuum pump.

Illus. 7-30. Attach the workpiece to the holding device and butt it against the starting pin.

Illus. 7-31. Feed the work into the cutter. Note that the shaping begins on edge grain, not end grain.

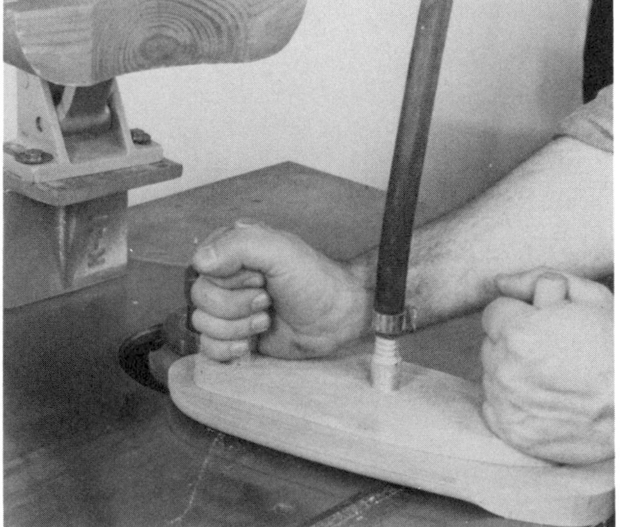

Illus. 7-32. Hold the stock against the rub collar while the shaping progresses.

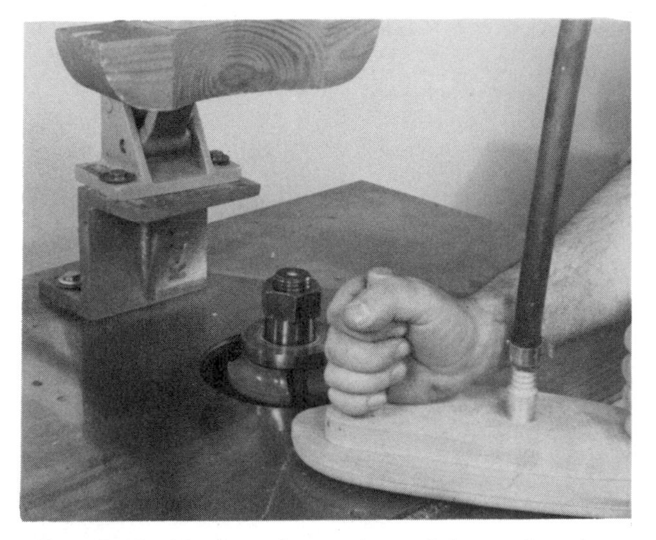

Illus. 7-33. After you have shaped the end grain on one end, pull the work away from the cutter so that you can reposition your hands.

Illus. 7-34. Here, one half of the workpiece's perimeter has been shaped.

Place a threaded rubber-hose connector on the top, at the center of the clamping board. Before doing this, however, drill a hole in the board to accept the threads. You can use various-size connectors. Use a length of three-ply black hose to link the hold-down to the vacuum pump. Make sure that you have used hose clamps to secure the hose connections. This will prevent leaks. Attach a relief valve either on the pump or conveniently to the hose so that you can easily release the vacuum when changing parts. The procedures for shaping with vacuum hold-downs are shown and discussed in Illus. 7-30–7-37.

Shaping with Jigs and Fixtures

USING THE MITRE GAUGE

A mitre gauge is normally supplied with the shaper. This device is similar to a mitre gauge used with a table saw. It slides parallel to the cutter in a slot cut in the top of the shaper. These slots may be angular or rectangular.

Mitre gauges are adjustable. Turning the mitre gauge enables you to make any angled cope or shaping cut easily. Various types of jigs and fixtures can be mounted to the mitre gauge.

The end grain of narrow pieces is normally cut using the mitre gauge. You can shape the end grain of pieces 12 inches wide or wider using the fences for control.

Mitre gauges are normally used with a stationary or split-fence system. When both fences are being used together, the mitre gauge slot should be parallel with the standard fence, so

Illus. 7-35. Again butt the work against the starting pin and feed it into the cutter. While it looks as if the cut is starting in the end grain, it is not. The end grain has already been shaped, so there is no chance of a kickback.

that mating parts such as those for cope joints can be accurately cut (Illus. 7-39 and 7-40).

Adjust the fence by using a combination square. Use the combination square as a depth gauge (Illus. 7-41). Hold its head against the edge of the mitre slot and adjust the distance between the mitre slot and the fence (Illus. 7-42). Always check measurements along the full length of the gauge slot.

If you are using a split fence, you must adjust both fences accurately to shape end grain. Never depend on holding the workpiece against the mitre gauge with only your hands.

To hold the stock securely, attach a piece of 80-grit garnet to the back surface of the gauge with double-face tape. You can also attach an auxiliary face to the mitre gauge (Illus. 7-44). These methods give you greater control and ensure that there will be less tear out.

Illus. 7-36. The work has been shaped completely. You can make another cut, if desired; just repeat the process.

Illus. 7-37. You can shape irregularly shaped raised panels using the vacuum hold-down.

Illus. 7-38. Raised panel doors are made with a set of door cutters. Usually the joints on the door frame are cut first.

Illus. 7-39. Once you have positioned the cope cutters on the arbor, adjust the fences.

Illus. 7-40. Use a straight edge to put the fences in a true plane.

Illus. 7-41. Use the combination square to adjust the outfeed fence uniformly from the fence.

Illus. 7-42. Adjust the infeed fence to the same relative position. The fence assembly is now parallel with the mitre slot.

Illus. 7-43. Once you have mounted the coped fences to the fence assembly, adjust them to the full depth of cut. Then back the infeed fence off slightly so that you can joint the entire edge away.

Illus. 7-44. Attach a wooden extension to the mitre gauge to compensate for tearout.

Illus. 7-45. This hold-down is used in conjunction with the mitre gauge.

Illus. 7-46. The mitre gauge and hold-down ensure that the work is held securely for end-grain coping.

In addition to the abrasive, attach a toggle clamp to the tongue of the mitre gauge. The abrasive and toggle clamp will control the movement of the workpiece (Illus. 7-45 and 7-46). The clamp will release just as quickly as your hands, but the toggle clamp is much safer. Shaper cutters tend to pull end grain into them as the cut is being made.

Make sure that the mitre gauge has no side play in the slot. If it does, the cut will be inaccurate, and there is a chance that a kickback could result. Make sure that the pivot bolt in the head of the mitre gauge is not loose. Tighten it if necessary.

Friction in the slot will eventually wear the tongue on the mitre gauge. This will produce a loosely fitting mitre gauge. To improve the fit of the gauge, use a center punch on the side of the tongue to produce dimples. These dimples will make the mitre gauge fit more tightly in the mitre slot.

USING STANDARD FENCES

Split fences usually come as a standard option with the shaper. They are made of metal and have removable wood face boards. The infeed and outfeed fences can be adjusted independently for the desired type of cut. Each side moves forward or rearward from the centerline of the cutter. (Note: On some shapers, there is only one adjustment knob for both fences. To move the outfeed fence, clamp the infeed fence down and turn the adjustment knob. If you wish to adjust the infeed fence, lock the outfeed fence and turn the adjustment knob.) The wood faces move towards or away from the cutter. The position of the split fences depends on the following: the diameter of the cutter, its profile, and the depth of cut being made.

When you are using the split or offset fence

system, you can joint some profiles during the shaping process. This is done by offsetting the fences slightly (Illus. 7-43). When offsetting the fences, adjust the outfeed fence first with a straightedge. Place the straightedge on the fences. It should extend to the deepest part of the cutter. Use the cutting edge of the cutter to make this adjustment.

Next, move the infeed fence back approximately 1/16 of an inch. This offset allows the cutter to joint the work as it is shaped. This jointing setup is designed for true stock only. Its purpose is to joint away small irregularities such as those caused by a saw blade.

You can use standard fences to make angular profile cuts by replacing their wood face boards with angular face boards. A rabbeted-edge board that has been cut at many diffferent angles provides an excellent hold-down for small workpieces. Wax all the wood face boards on the fence system generously to reduce friction.

Standard, flat 3/4-inch face boards should be made of a good grade of sheet stock. These materials move less with changes in climate than hardwood boards.

If you are using fences for a production setup, glue a plastic laminate to the fence. This will reduce wear on the fence. Making a cope cut on the fence boards that is the shape of the cutter will help to bring the fences closer to the cutter. The closer the fences are to the cutter, the easier it is to control the workpiece.

Production Shaping

Production shaping is the process of shaping multiple pieces, either freehand, feeding them with fences, or by power feeding them. All of these methods use the same setup over and over to create hundreds of finished pieces.

Before setting the shaper up for production-shaping, consider the following points carefully: 1 The material used for the jigs and fixtures is important. Select materials that will show the least amount of friction and wear over repeated use (Illus. 7-47). 2 Evaluate the cutters to determine which one will last the longest. In some cases, this may mean cutters made of carbide or Tantung™. In other cases, cutters made of tool steel will be adequate. The type of wood being shaped will help you decide which cutters will work best. 3 Determine if the horsepower of the motor on your shaper can handle long cutting runs. The depth of cut can help determine this.

A three-horsepower motor will provide enough power for most multiple cutting runs. However, constantly overheating the motor will create problems. An application of paraffin wax to the table and fences reduces friction and stress on the motor. It also reduces the temperature the cutter cuts at. Also, hone the cutter now and

Illus. 7-47. Production jigs like this Panelcrafter allow you to perform a number of production jobs safely without getting your hands near the cutterhead.

Illus. 7-48. A backing board has been mounted on the Panelcrafter. This backing board will control tearout.

then to increase its tool life and improve the quality of the cut it produces.

When shaping multiple pieces freehand, clean the rub collar periodically to avoid pitch buildup. An accumulation of pitch can create marks or blackening of the wood where the rub collar contacts the wood.

During production runs, periodically check the shaper for vibration. Vibration can loosen shaper settings when they are used extensively. Spindle gib screws can sometimes loosen, and fences can also move due to vibration. Get in the habit of checking these parts regularly during a production run.

Shaping Doors

The shaper can be used to shape doors of all types. The most common types of doors made on the shaper are cabinet doors and house doors.

Cabinet doors are usually ¾–1¼ inches thick. Interior house doors are 1⅜ inches thick, and exterior house doors are 1¾ inches thick.

The most common doors made on the shaper consist of a frame and a series of wood panels or glass panes. Some doors have a mix of wood panels and glass panes.

The frame of most shaper-made doors is joined with cope-and-stick joinery. A cope-and-stick joint is a two-part mating joint. The cope is cut on the end grain of the work. The stick is cut on the edge grain of the work.

Cope-and-stick joints are commonly used in interior and exterior door construction and for panelled walls. They are also used in furniture carcass construction. The two parts of the joint fit snugly together and provide a groove or rabbet that accommodates glass, or a raised or plain panel.

Two separate sets of cutters are used to make a cope-and-stick joint. One set of cutters shapes the cope, while the other set shapes the stick. Some of the basic coping techniques are presented in Illus. 7-38–7-46.

CUTTING COPE-AND-STICK JOINERY

To set up the shaper, start with the stick or female cutters. The set of cutters should be the correct size for the thickness of stock you are using. Cut a sample workpiece. Use stock that is the same thickness as the wood you will be using for the job.

Adjust the height of the ¼-inch groove cutter so that it is more than ⅛ inch from either face of the workpiece. Position the infeed fence so

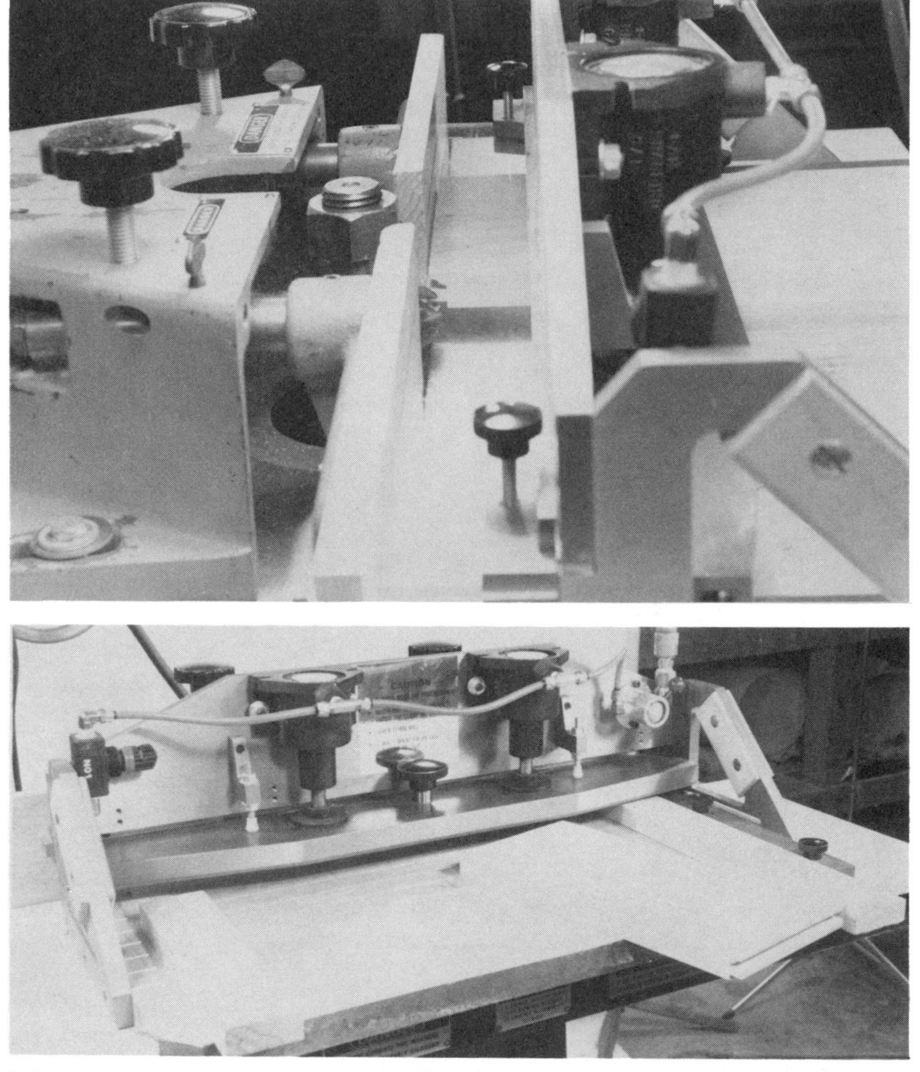

Illus. 7-49. The air clamps are holding the work for shaping. Some air escapes from the bottom of the jig to make it easier to control.

Illus. 7-50. When the air clamp is released, the work can be reversed easily. Be sure to keep the good face up (or down) when you shape the opposite end.

that it is 1/16 inch back of the inside cutting edge. Adjust the outfeed fence so that it is in line with the inside cutting edge.

Once you have cut a sample stick on the edge of a setup piece, mount the coping cutter set. Cutting a stick sample first allows you to space the cope cutters (using the shim stock provided by the manufacturer). This results in a perfect mating joint. The cope-and-stick joint should have a snug fit. However, if it is too tight, the glue will not bond correctly.

The height of the cope cutter is taken from the stick sample (Illus. 7-51). Position the groove cutter using the setup piece. The depth of the cut will be the same.

Since cope cuts are made on the end grain of the workpiece, use a mitre fence or a special jig. Use the mitre slot to hold the workpiece square

Illus. 7-51. One of the easiest ways to adjust the height of a cutter is with the mating part. Align the groove cutter with the tongue.

Illus. 7-52. A power feeder is being used to cut the stick or female part of the joint.

with the fence. Make sure that the fence is running parallel to the mitre slot. Check both the infeed and outfeed sides.

Adjust the mitre gauge so that it is square with the straight fence. Attach a backing board to the mitre gauge. Position the backing board next to the straight fence and clamp it to the mitre gauge. Shape the backing board first to produce a sample cut to match with the stick sample.

Although backing boards are used for tearout when end grain is being cut through, add enough width to the cope stock so that it can be jointed off when the stick is shaped. After you have coped all the end grain, edge-shape the mating profiles (Illus. 7-52). Remove the cutters

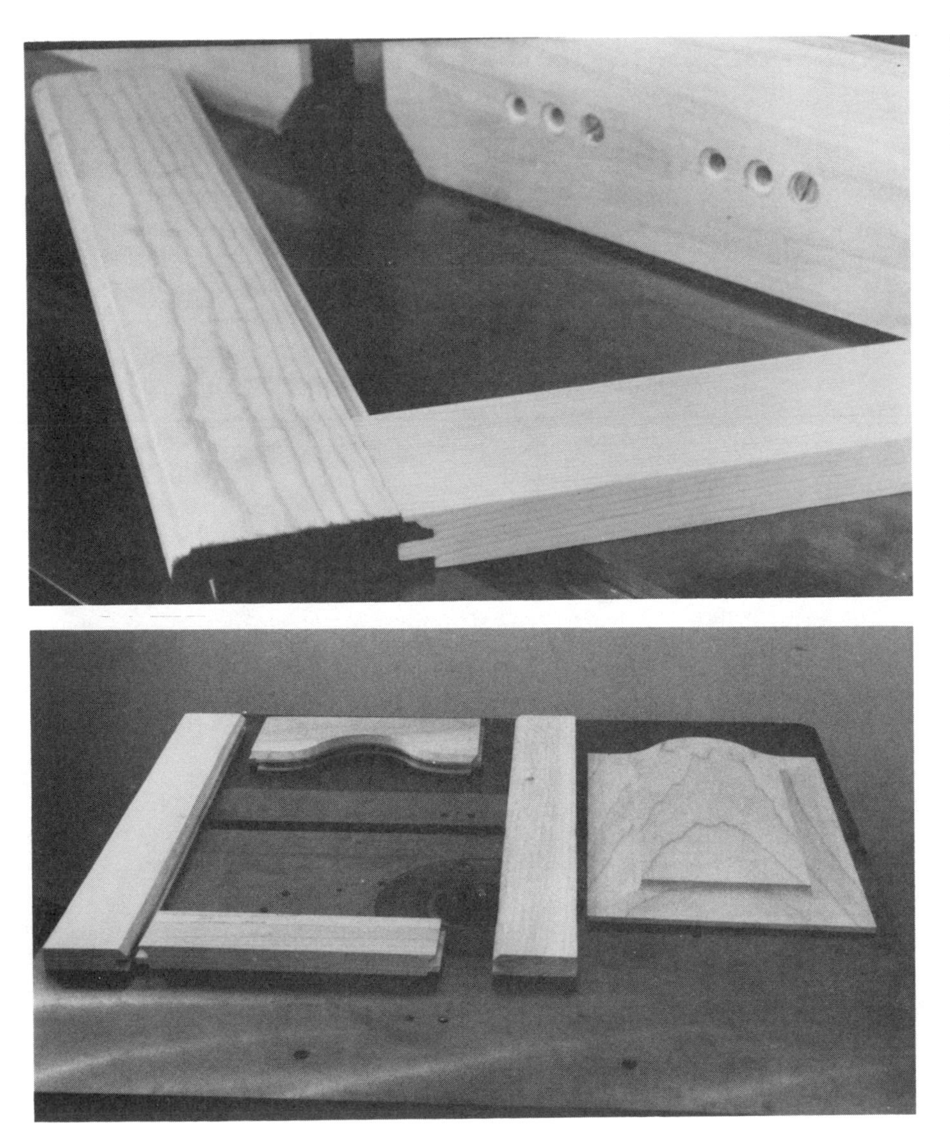

Illus. 7-53. The cope-and-stick joint should be snug, but not so tight that glue cannot be used.

Illus. 7-54. A crown rail will require a different shaping technique due to its irregular edge.

and replace them with the mating cutter set. Adjust their height again with the tongue-and-groove alignment. Their depth stays the same, so the position of the fence remains the same.

Shape straight edges using the fence and standard hold-downs. Test the fit between the cope-and-stick cut. If it is too tight, add shim stock between the cutters to improve the fit (Illus. 7-53). Consult the manufacturer's instructions that were furnished with the cutters.

Rails and stiles with irregular edges can be shaped with the stick cutters (Illus. 7-54). This is done using jigs cut to the profile desired (Illus. 7-55). Fasten the work to the jig. To minimize damage to the wood, use pneumatic or vacuum clamping. Remove the dust hood and straight fences from the shaper table surface.

Position the starting pin as close as possible to the cutterhead. If the jig is longer than the rail, you will not have to use a starting pin. Place the rub collars used with the cutter sets above the cutters, for safety. If you are not using backing boards for the cope cuts, make sure that the workpiece is wide enough to be trimmed during the stick-shaping operation.

Determine the length of the rails with the aid of the manufacturer's cutting details. There is no need to add extra length for jointing because the jigs take care of the depth of the cut.

Cope-and-stick cutters can shape stock with its good face up or down (Illus. 7-56–7-61). For production work, many woodworkers prefer to

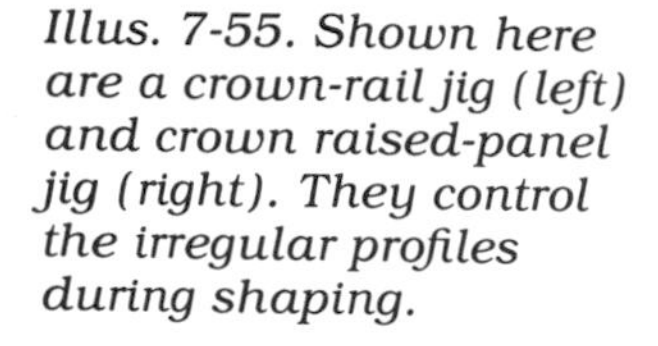

Illus. 7-55. Shown here are a crown-rail jig (left) and crown raised-panel jig (right). They control the irregular profiles during shaping.

Illus. 7-56. Cut the cope joint on both ends and saw away the arch before shaping the rails.

Illus. 7-57. Line the tongue part of the cope joint with the groove-cutting part of the shaper head. This ensures proper fit and alignment. No starting pin is needed because the jig engages with the rub collar before the work touches the cutterhead.

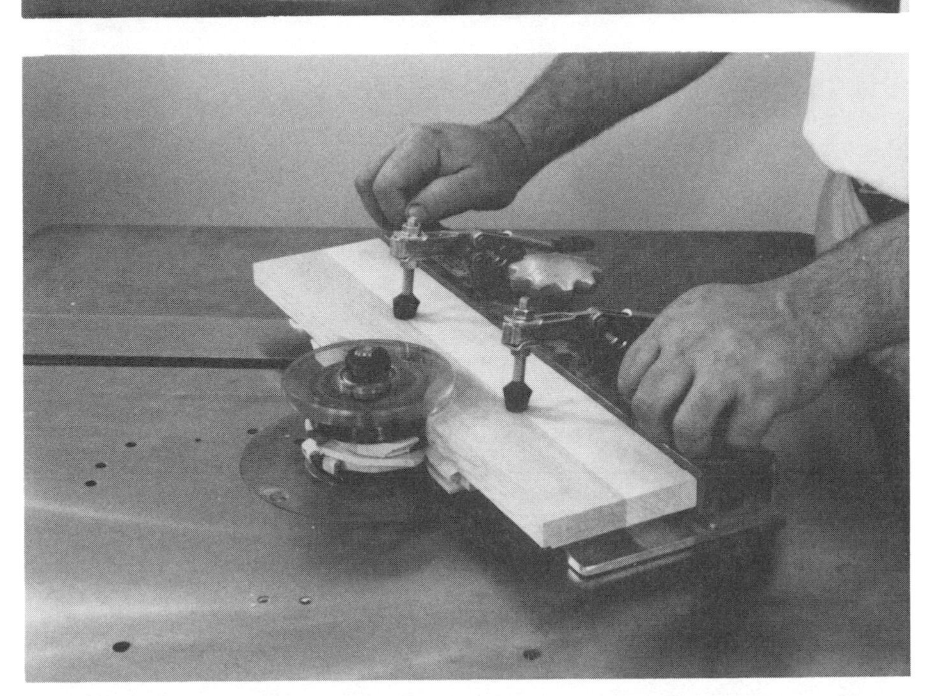

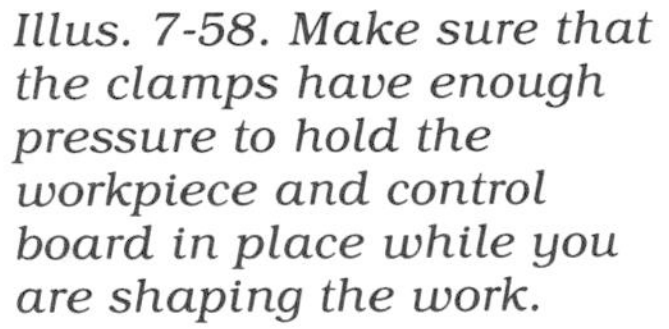

Illus. 7-58. Make sure that the clamps have enough pressure to hold the workpiece and control board in place while you are shaping the work.

Illus. 7-59. When you have shaped the rail, back it away from the cutter.

Illus. 7-60. This setup uses a two-wing cutter. A two-wing cutter reduces burning because there is less friction between the wood and cutter. (Photo courtesy of L. A. Weaver)

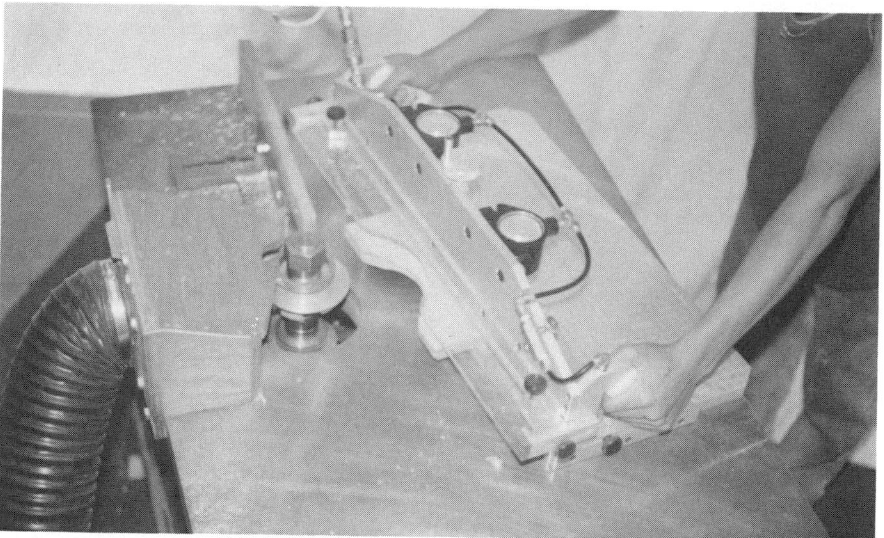

Illus. 7-61. This Shapermaster™ is being used with a template to shape a crown rail. (Photo courtesy of L. A. Weaver)

Illus. 7-62. This wood panel was raised on the shaper. A horizontal or vertical cutter can be used to raise a panel.

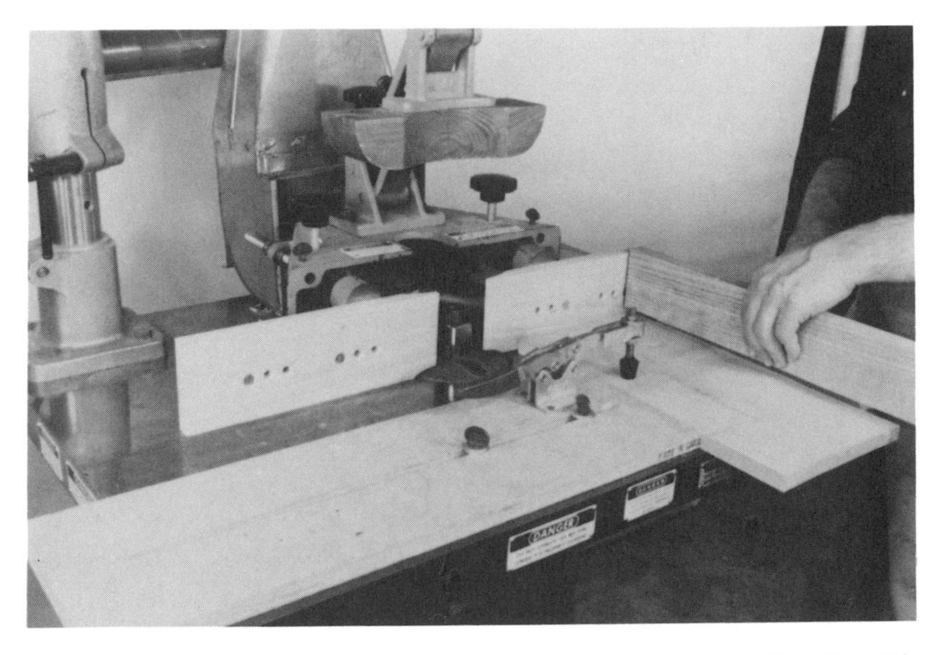

Illus. 7-63. Adjust the fences close to the cutter, and attach a special setup to the mitre gauge to control end grain. A hold-down helps ensure safe and accurate work.

shape stock with its good face down. This ensures that the front of the door lines up perfectly even when there is some deviation in the thickness of the stock. You can remove the irregularity on the back of the door with a wide belt sander after assembling the door.

For custom work, the approach is not as critical, provided that all stock is uniform in thickness. A final light planing of all parts could be the first step in quality door production.

Shaping Raised Panels

Special shaper cutters are used to raise wood panels (Illus. 7-62). These cutters are either mounted horizontally or vertically. The vertical cutter tends to tear the end grain. The horizontal cutter works best on crown-top doors.

Vertical cutters should be used on shapers with at least a 3/4-inch spindle. Smaller spindles can deflect as the work is fed into the cutter. This can cause poor cutting. Horizontal cutters have a lower profile and do not deflect as easily, but they also work best on spindles at least 3/4 inch in diameter.

Some woodworkers control the size of the chips by using a cutter with a limiting shoulder or small gullet. These cutters are the safest to use because they are least likely to kick back. See Chapter 3 for a discussion of shaper cutters.

Panels are more difficult to shape if they are not perfectly square and flat. Save your best wood for panel shaping. Check it before shaping for cupping, twisting, knots and checks. These defects can cause the panel to disintegrate during shaping. Depending on the material used, the panel size and the application, the panel should have at least 1/16 inch clearance on all four sides. This extra clearance allows for expansion in the frame.

Check the thickness of the panel stock. It should be appropriate for the shaper cutter you intend to use. After raising the panel, you should have a 1/4-inch tongue that is 3/8 inch in from the edges of the panel. This tongue will match the 1/4-inch groove cut in the frame.

Adjust the depth of cut for the raised panel. When using the horizontal cutter, adjust the straight or split fence to the maximum depth of the cutter knives (Illus. 7-63). When using the vertical cutter for raised edge work, adjust the depth of the cutter by raising or lowering the spindle. *When changing the height of the cutter, always turn off the power.*

The depth of the profile must also be adjusted. Adjust the horizontal cutter by raising or lowering the spindle. The depth of cut will depend on the type of wood used, the profile of the cutter,

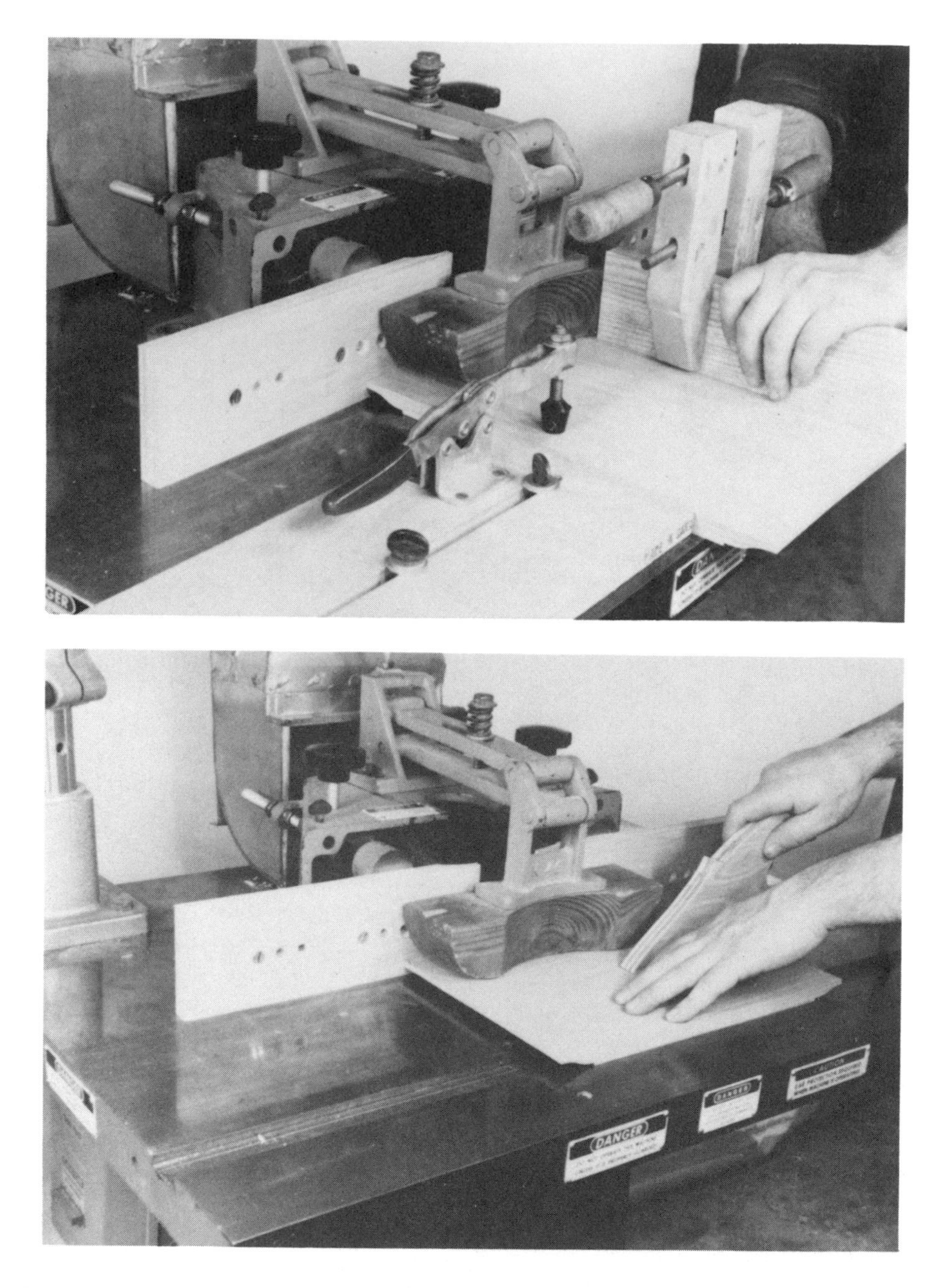

Illus. 7-64. The clamping device keeps the end grain from being pulled into the cutter while it is being shaped.

Illus. 7-65. After you have shaped all end grain, the edge grain can be shaped.

and the horsepower of the shaper. Most raised panels require at least two passes. Do not raise the spindle to the finished depth for the first cut. It is much better (and safer) to take two or three light cuts. Remember, you will be cutting end grain one half of the time. End grain is much harder on cutting tools (Illus. 7-64 and 7-65). Heavy cuts will generate heat and burn end grain.

When using the vertical cutter, adjust the depth of the profile by moving the straight fence in or out, parallel with the cutter, or by raising and lowering the cutter. Again, do not take the finish pass with the first cut. Drop the spindle and work your way up to the finished size with two or three cuts.

Move the split fence as close to the cutter as possible when using the vertical raised panel cutter. Remember that you will be adjusting the cutting depth again for the second and third cuts. Readjust the fences each time you elevate the spindle. Cope the fences so that they fit closely around the cutter. A close fit gives you greater control over the work and makes the operation safer.

For a safer operation of the horizontal cutter, attach a rub collar to the spindle. This will prevent the cutter from grabbing the work and spoiling the edge or end profile or causing a kickback.

Make sure that all guards are in place. Adjust the hold-downs for the correct holding pressure (Illus. 7-66–7-72). Wax the table and hold-downs with a bar of paraffin so that the stock can be moved more easily.

Consider the size of the panel before shaping. Use a mitre gauge when shaping the end grain on panel stock that is less than 12 inches wide.

Illus. 7-66. This special raised-panel setup features guarding and holding capabilities.

Illus. 7-67. A wooden dust-collection hood was made to make the operation safer. The triangular blocks push the chips to the chute.

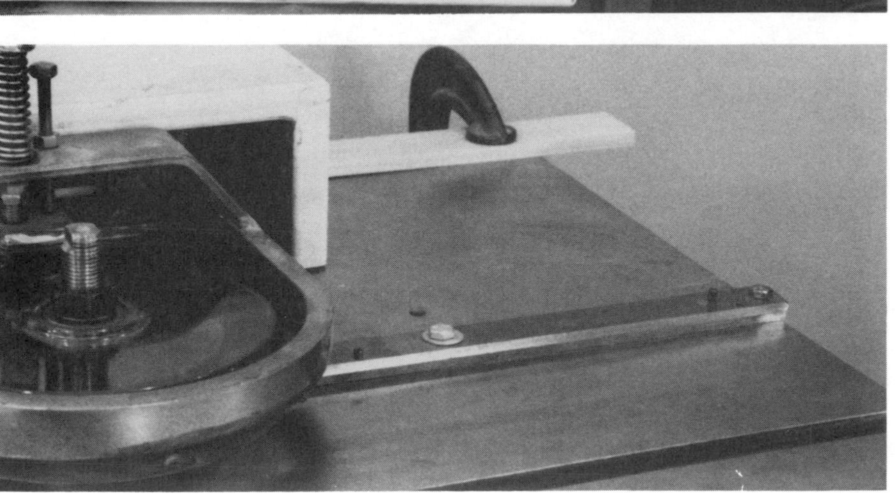

Illus. 7-68. A special low fence is bolted to the shaper table for this raised panel setup. The fence is about the same height as the tongue on the raised panel.

Illus. 7-69. This crank controls the tension on the hold-down. Too much tension can damage the face of the panel.

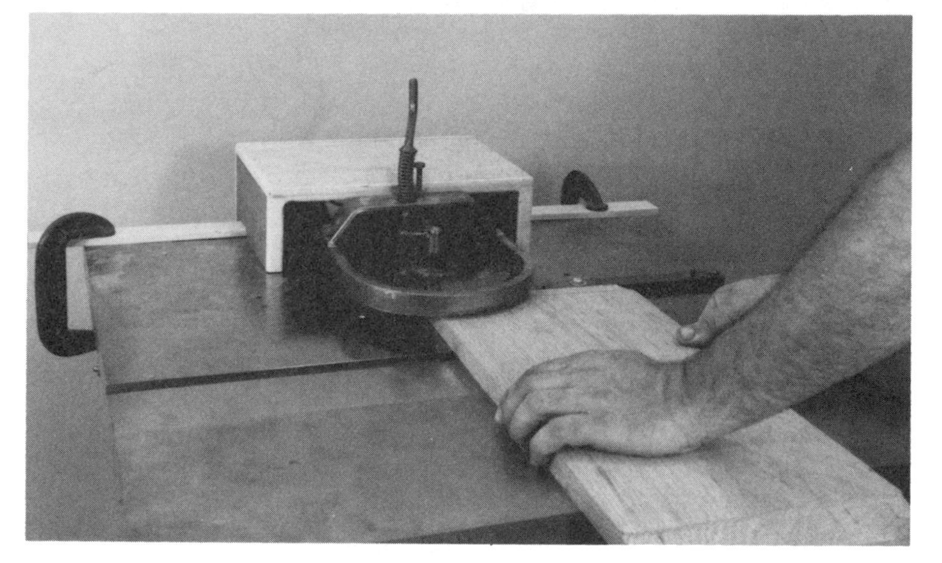

Illus. 7-70. The fence and the rub collar are in the same plane. This allows the rub collar to control the depth of cut after the work leaves the fence.

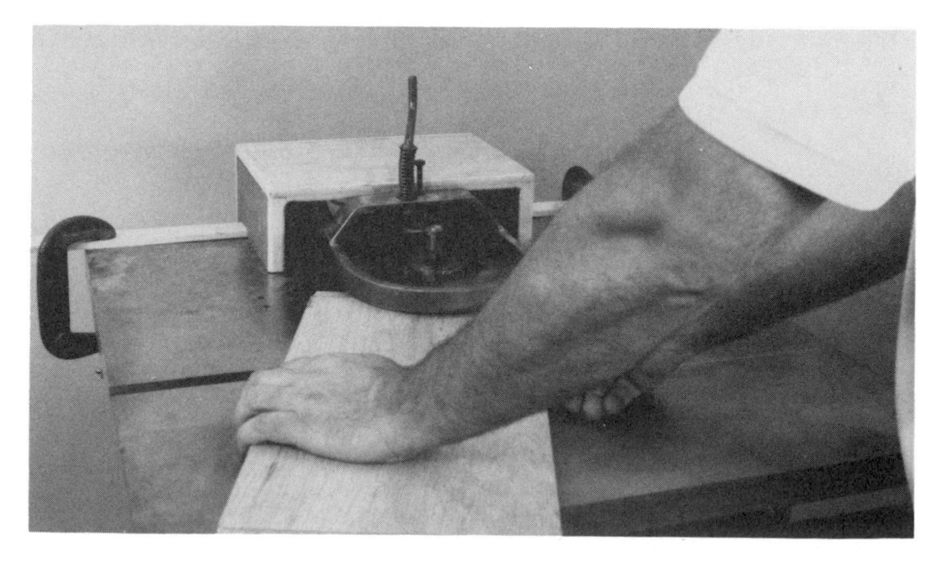

Illus. 7-71. The panel can pivot on the outfeed side of this setup. This allows crown panels to be shaped without a jig.

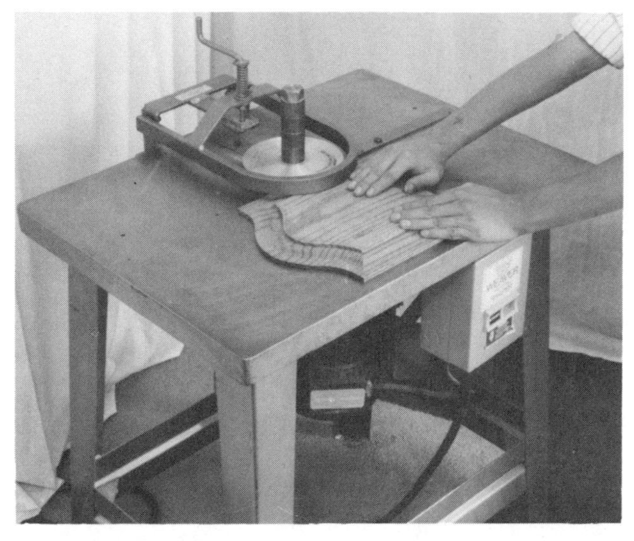

Illus. 7-72. Irregular profiles, end grain, and edge grain can all be shaped with this setup. (Photo courtesy of L. A. Weaver)

Shallow passes are best. Shape all edges of the panel.

Start shaping with end grain, and end with edge grain. End grain usually chips at the end of the cut. Shape the end grain at a slightly faster feed speed. This will help prevent burning. Slow down at the end of each cut.

As mentioned, when changing the height of the cutter, *always turn off the power.* Always test the setup after adjustments have been made. If the cut seems too deep, readjust the shaper.

Some small or irregularly shaped panels (such as crown panels) cannot be shaped with the vertical, raised-panel cutter. Shape these panels using any horizontal cutter. Control the panel stock with the help of jigs and fixtures or vacuum hold-downs.

Illus. 7-73. A rub-collar bearing and straight cutter can be used to joint an irregular profile to the desired shape.

Illus. 7-74. This edge-jointing operation uses a dead collar under the cutter. A dead collar remains stationary while the cutter turns. (Photo courtesy of L. A. Weaver)

Irregularly shaped workpieces are guided by a starting pin, as outlined earlier. The starting pin minimizes the chance of kickback.

The workpiece can be controlled with just a rub collar or a rub collar and a profile jig (Illus. 7-73–7-82). If using a profile jig, cut it to the desired shape and clamp the workpiece to the jig. The jig then rides against the rub collar.

Vacuum hold-downs can also be designed to ride against the rub collar (Illus. 7-83 and 7-84). Vacuum hold-downs minimize hand involvement and eliminate the need for a separate jig for profile shaping.

When shaping irregular panels, start with side grain and proceed in the counterclockwise rotation if the machine rotation is counterclockwise. Feed the stock into the cutter. A single stroke at a good feed rate will minimize burning. Once the wood has been engaged with the cutter, continue cutting straight away from the

Illus. 7-75. The jig, cutter, and rub collar are positioned for shaping. The guard above the cutter makes the operation safer. No starting pin is needed because the jig is wider than the panel.

Illus. 7-76. Hold the jig firmly while making the cut. Make sure that all the clamps are adjusted to their maximum holding pressure.

Illus. 7-77. Inspect the work. If burning occurs, take a lighter cut, increase feed speed, or use a cutterhead with fewer wings or cutters.

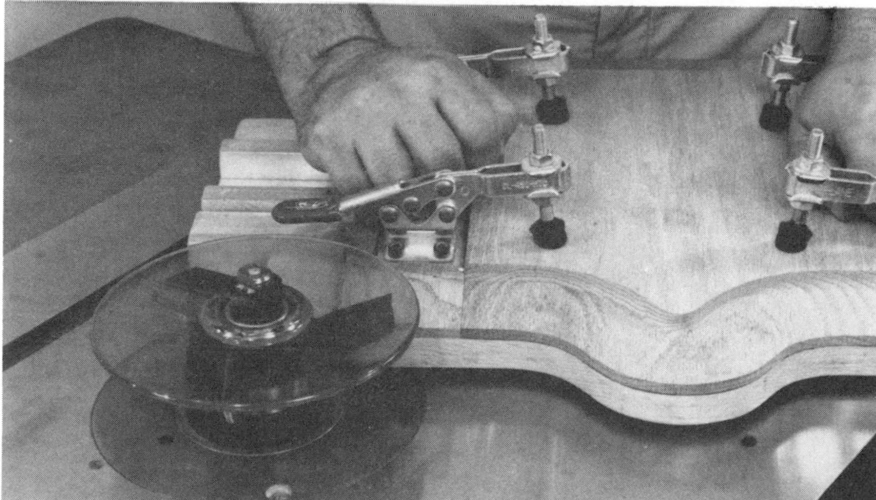

Illus. 7-78. Take the work completely through the cutter before backing it away from the rub collar.

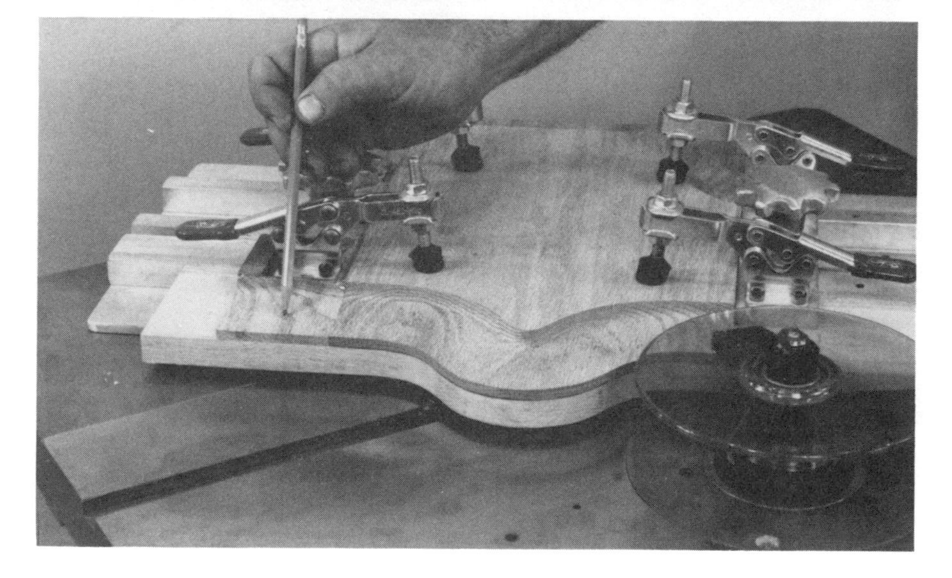

Illus. 7-79. A special block is attached to the jig to compensate for grain tear-out when shaping the end grain.

Illus. 7-80. You can position a straight-edge pattern on the same jig to shape straight end grain.

Illus. 7-81. Tear-out is again controlled by the special block mounted on the jig. Always complete the cut with the pattern against the rub collar.

Illus. 7-82. This completed crown-top raised panel is ready to be assembled to the door.

Illus. 7-83. This irregularly shaped raised panel requires the use of a vacuum chuck.

Illus. 7-84. The rub collar rides along the edge of the work while the vacuum chuck holds it securely.

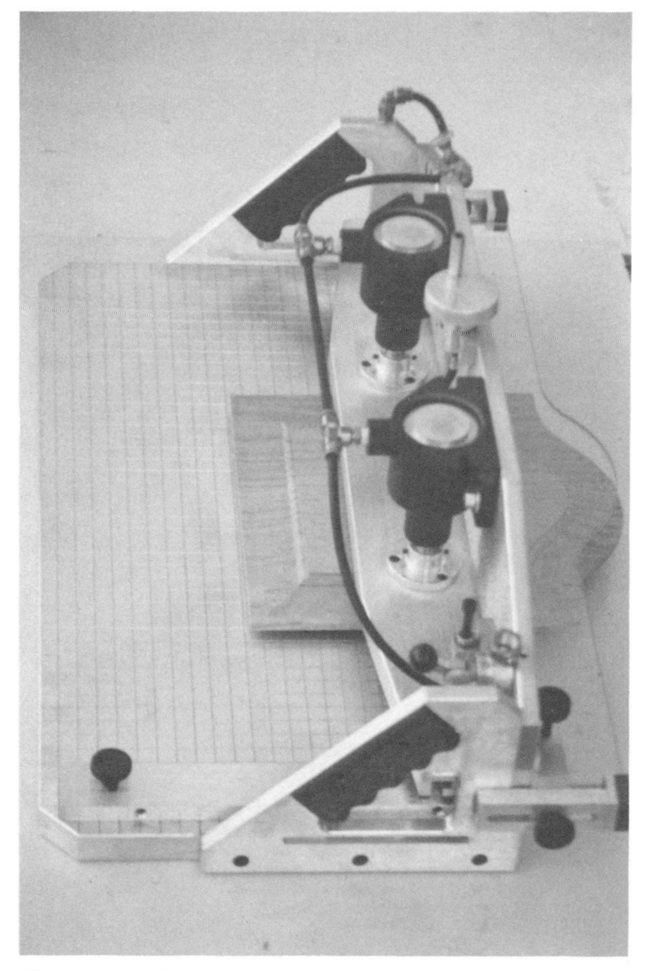

Illus. 7-85. This pneumatic jig holds the work securely while it is being shaped. (Photo courtesy of L. A. Weaver)

cutter. When starting again, start back from the end of the cut previously at a distance that's at least the diameter of the cutter. This will prevent grabbing of the wood when it comes in contact with the half-shaped area.

PNEUMATIC AND POWER-FEED APPLICATIONS

Holding panels with pneumatic jigs while shaping can also make the job safer and more efficient (Illus. 7-85 and 7-86). The air pressure holds the panel securely and allows the jig to float on the table (Illus. 7-87). The plastic template rides on the rub collar to control the pattern being shaped. The operator's hands are kept well away from the cutter.

Power-feed units can be used to feed panels into horizontal panel-raising cutters (Illus. 7-88). Rectangular and square panels can be fed through the cutter at a feed rate designed for the wood used (Illus. 7-89).

To set the shaper up for this operation, first adjust the fences as described previously. Set the power feeder according to the owner's manual; adjust its height and taper it towards the fence. The power feeder provides a chatter-free cut (Illus. 7-90).

A power feeder is usually set up only when there are multiple panels. On a production run,

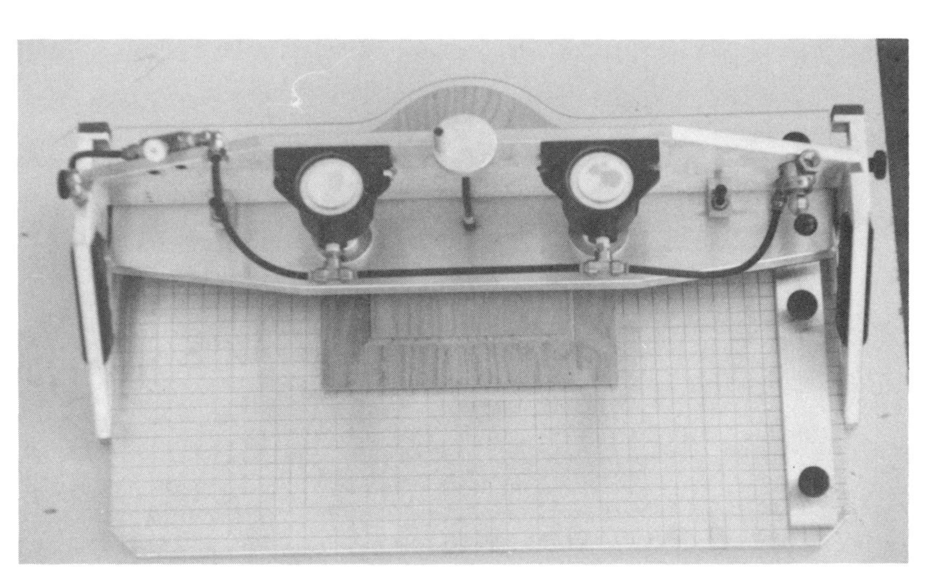

Illus. 7-86. The handles provide control and keep the operator's hands well away from the cutter. (Photo courtesy of L. A. Weaver)

Illus. 7-87. The plastic template controls the path of the cut, and the air pressure holds the panel securely and allows it to float on the shaper table. (Photo courtesy of L. A. Weaver)

Illus. 7-88. Power-feed units can be used to shape raised panels with straight edges and ends. The fences must be positioned the same way they would be positioned for hand feeding.

Illus. 7-89. Adjust the feed rate according to the wood species, the depth of cut, and the horsepower of the shaper. The power feeder should be positioned to push the stock against the fence.

Illus. 7-90. Power feeders produce a chatter-free cut and reduce wear on the tooling.

the operator can safely run numerous panels with a high degree of finished quality and dimensional uniformity (Illus. 7-91).

When a door consists of several panels, mullions are shaped to go in between the rails and stiles. These mullions require cope-and-stick cuts, as well. For best results, use jigs and fixtures for control. These smaller parts are difficult to shape without a jig or fixture to hold them while they are being cut.

Power feeders can also be used for the stick cuts made on mullions. If the mullions are quite short, it is best to shape a long piece and cut it into smaller parts. You can then shape (or cope) the end grain with the aid of a jig attached to the mitre gauge. Be sure to clamp the parts securely to keep them from being pulled into the cutter.

SHAPING RAISED PANELS WITH A VERTICAL CUTTER

While there is more deflection in vertical raised-panel cutters, there is less weight. This means less energy is used to turn the cutter. The result

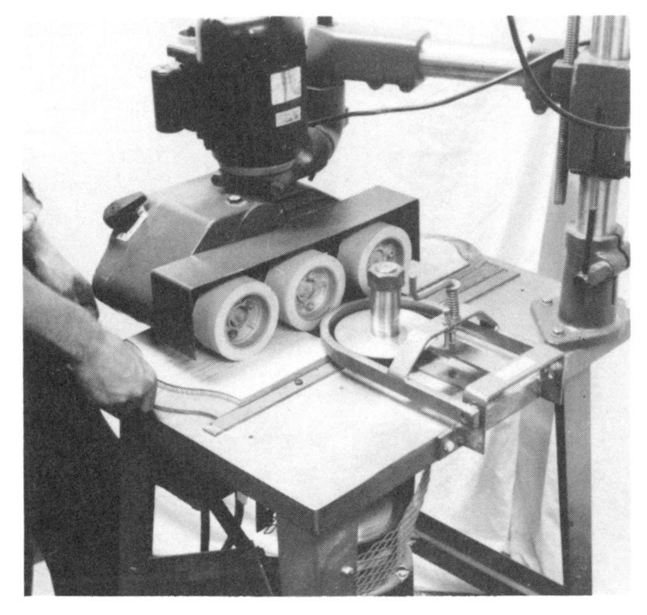

Illus. 7-91. After you have shaped the crown, shape the straight edges using a power feeder. (Photo courtesy of L. A. Weaver)

Illus. 7-92. When shaping raised panels with a vertical cutter, cope a single fence to its profile. Note the relief above the coped area for the spindle.

is more energy left for cutting wood. This is important when you are raising panels on light-duty shapers. The smaller diameter means reduced peripheral speed. Lower peripheral speed reduces the chance of burning, even at slow feed rates.

For best results, replace the split fences with a single-piece fence (Illus. 7-92 and 7-93). Cope this fence closely to the cutter profile. This will help reduce tearout while you are shaping.

Mount the cutter and position the fence for a light cut (Illus. 7-94). Use a featherboard to hold the stock against the fence (Illus. 7-95). The featherboard also acts as a barrier between the operator and the cutter.

Once everything is locked securely, fit a push block to the fence (Illus. 7-96). To do this, rabbet it to the fence (Illus. 7-97). It will slide along the top edge of the fence. Position the stock for end-grain shaping, and use the push block to control the shaping (Illus. 7-98).

Once you have shaped the ends of all the pan-

Illus. 7-93. Mount the single fence as a replacement for the split fences. Make sure that the cope is lined up with the cutter before attaching it to the fence assembly.

Illus. 7-94. Adjust the depth of cut by moving the fence or the cutter. Make the setup for a light cut.

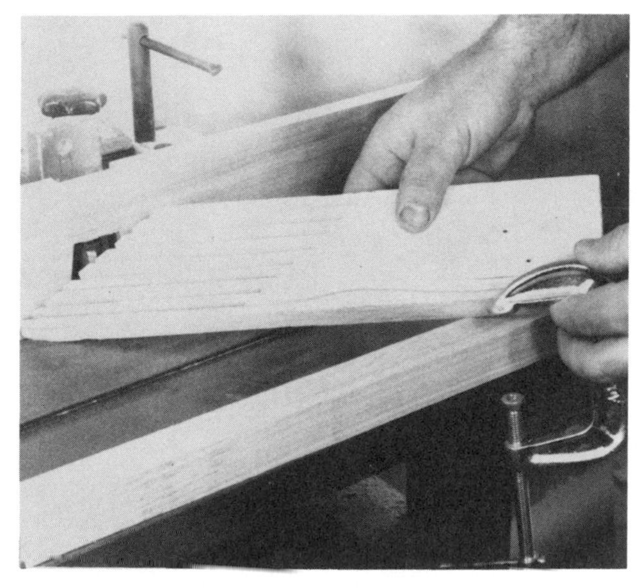

Illus. 7-95. Clamp a featherboard to the table. This will help keep the work against the fence during shaping. It will also act as a barrier between you and the cutter.

els, shape the edges. Readjust the depth of cut and replace the featherboard with a push stick (Illus. 7-99). A push stick is used when edge grain is being shaped.

Some curved-top panels can be shaped with a vertical panel-raising cutter if a curved jig is clamped to the fence (Illus. 7-100). The curve of the jig is the same curve as the one on the panel (Illus. 7-101). Clamp the jig to the fence well away from where the shaping is to occur (Illus. 7-102). Put the head of the clamp on the fence side to prevent it from getting in the way. Cut a recess in the jig near the cutter. Make sure that it is positioned correctly (Illus. 7-103) and that you handle it carefully.

Clamp a featherboard to the shaper table and position the work for shaping (Illus. 7-104). Turn the shaper on and guide the work into the cutter (Illus. 7-105). Keep the stock on the jig and the fence while shaping (Illus. 7-106). In-

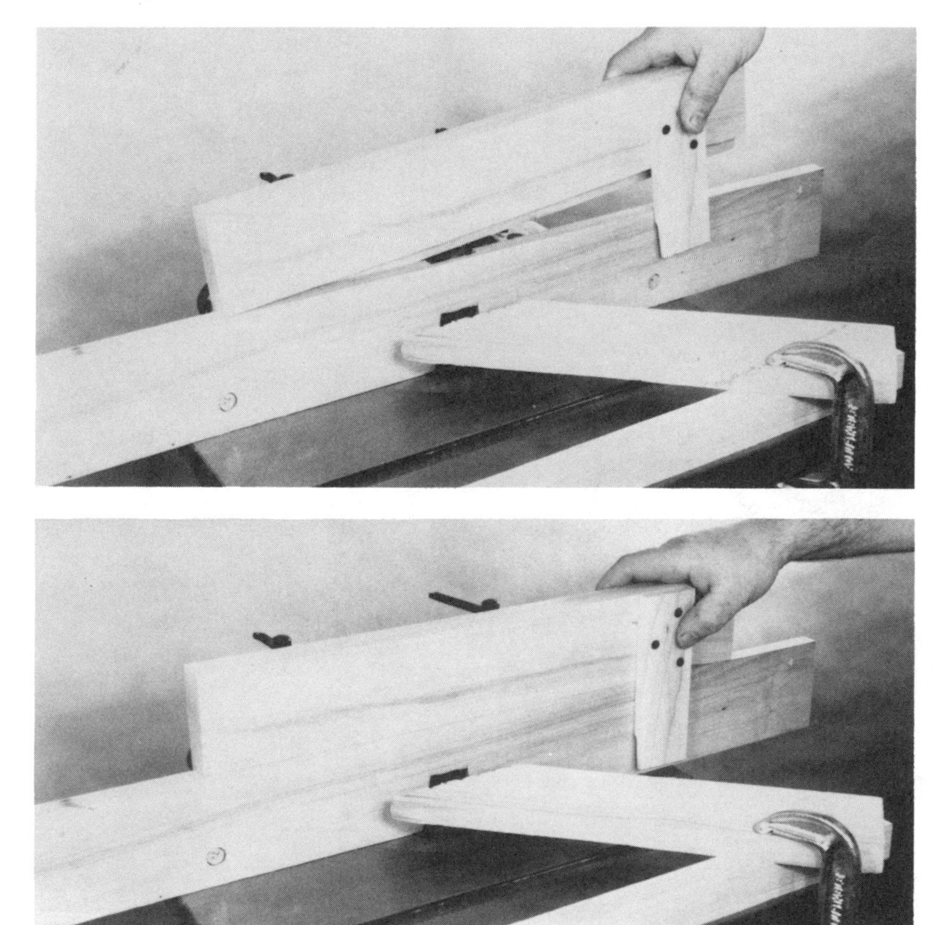

Illus. 7-96. Fit a push block to the fence. It will be used to guide the work during end-grain panel raising.

Illus. 7-97. Rabbet the push block so that it will slide along the top edge of the fence.

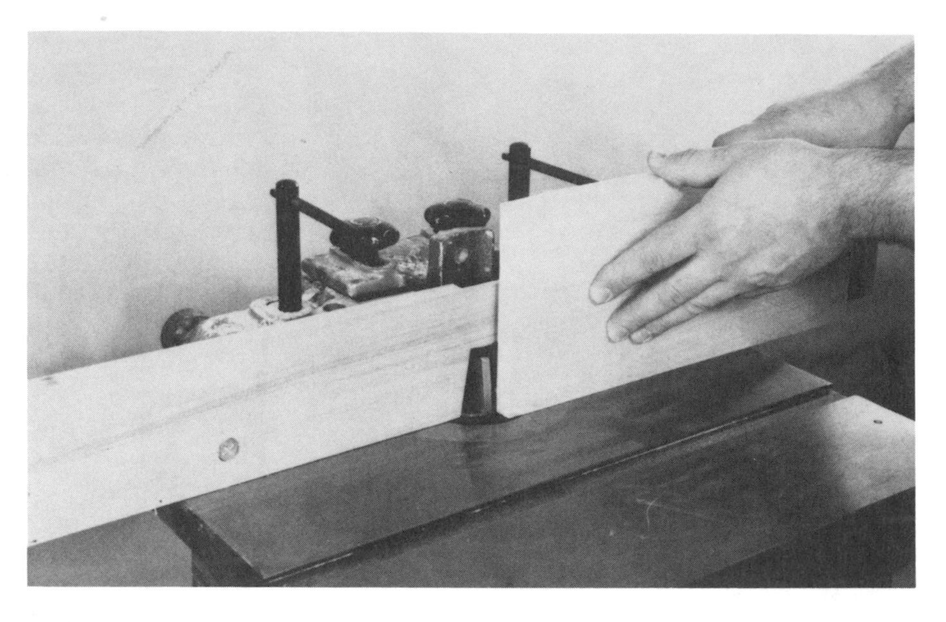

Illus. 7-98. Position the work against the table and push block. Slowly guide the work across the cutter. There is less chance of burning because the tip speed of the cutter is much slower than that of a horizontal panel-raising cutter.

Illus. 7-99. Use a push stick and a featherboard to control edge-grain panel raising. Re-adjust the depth of cut after you have shaped all the end grain.

Illus. 7-100. If you clamp a curved jig to the fence, you can use a vertical panel-raising cutter to shape curved panels. The curve of the jig varies according to the panel.

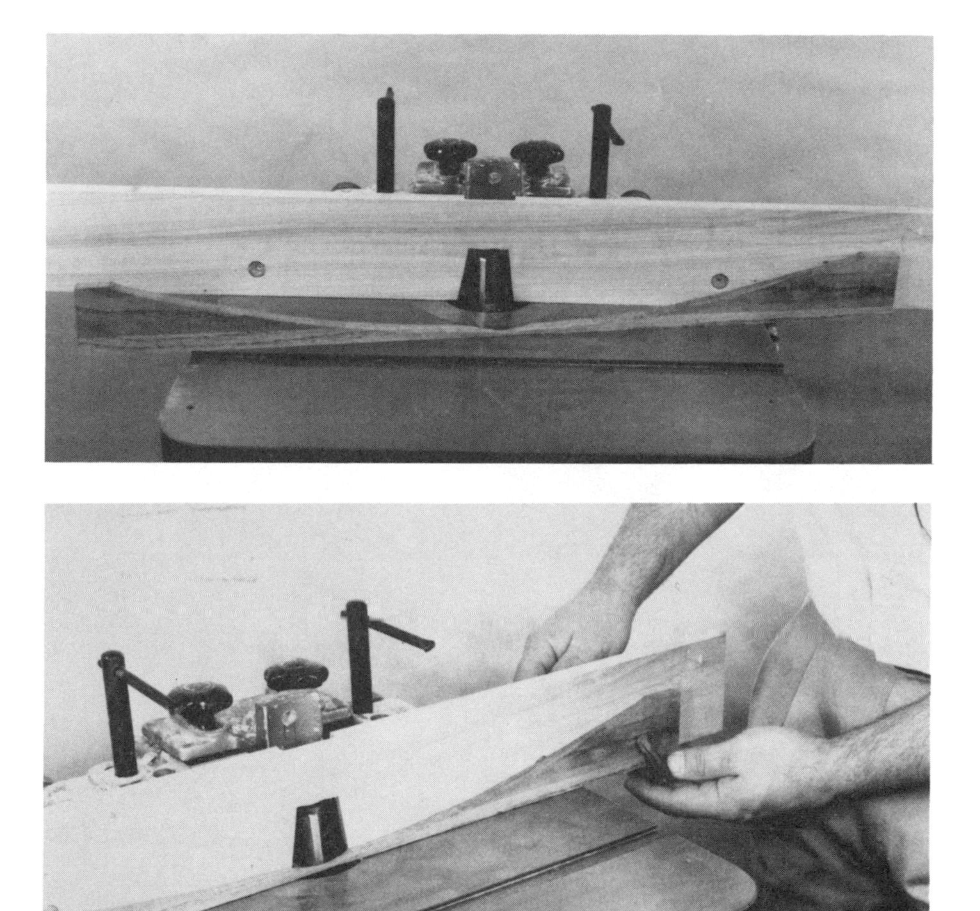

Illus. 7-101. This jig guides the panel through a smooth arc while it is being shaped.

Illus. 7-102. Clamp the curved jig to the fence. Reduce any obstructions while shaping by putting the head of the clamp on the fence side.

Illus. 7-103. Make sure that the cutout in the curved jig lines up with the opening in the fence.

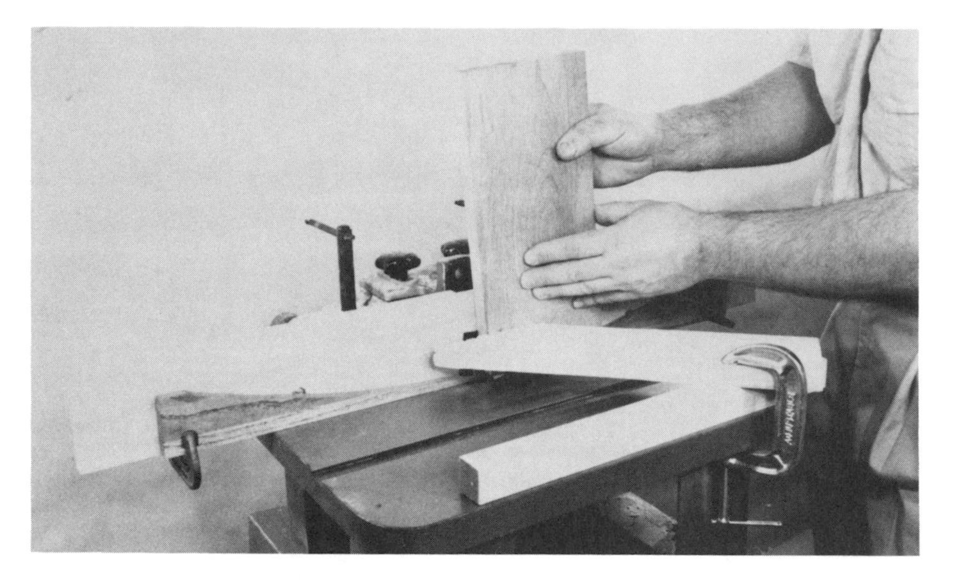

Illus. 7-104. Clamp a featherboard to the table and position your work on the infeed side of the cutter.

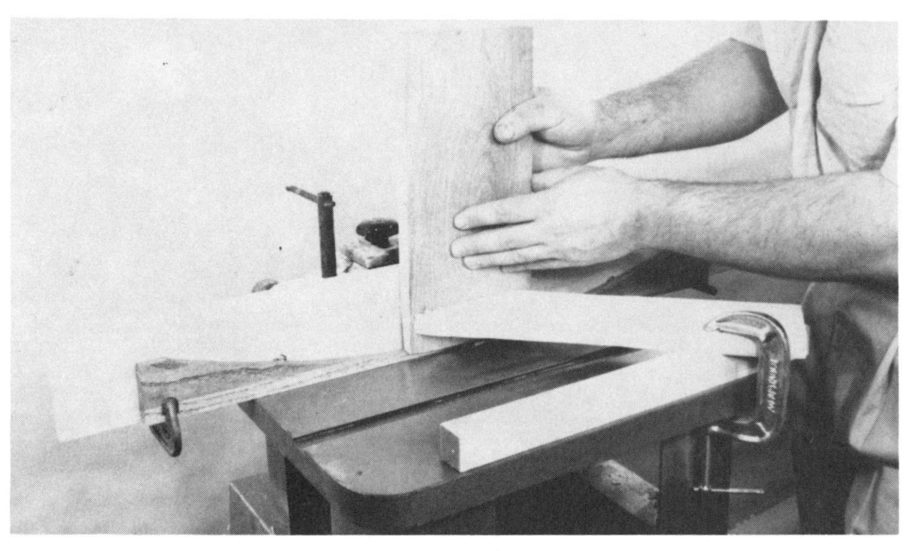

Illus. 7-105. Turn on the shaper and guide the work into the cutter. Keep the work on the fence and jig throughout the cut.

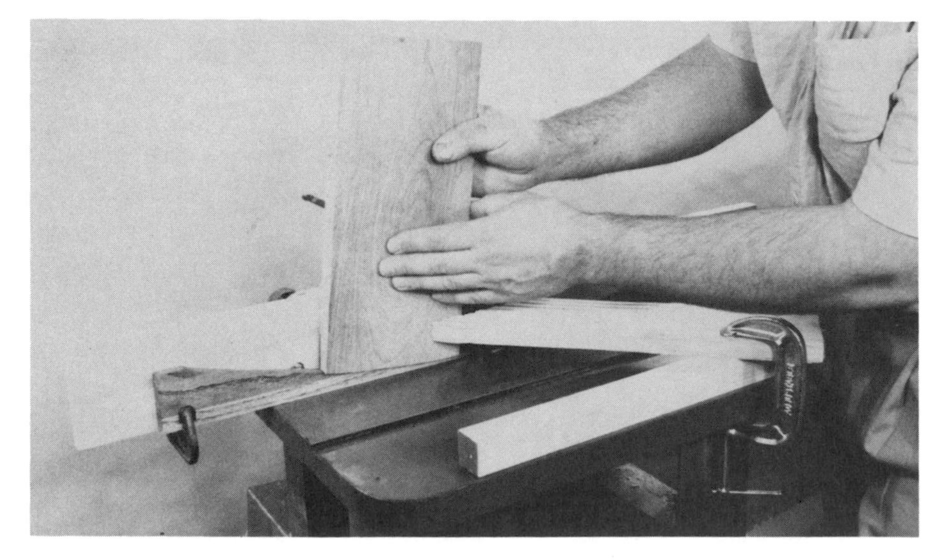

Illus. 7-106. A sweeping action is needed during this operation. Remember, the curve of the jig must be the same as the curve of the panel.

Illus. 7-107. Inspect the cut. Re-adjust the fence or cutter to make the cut deeper.

spect the cut (Illus. 7-107). If the cut is not deep enough, move the fence back or raise the cutter.

After you have raised all the curved ends (Illus. 7-108), remove the curved jig and shape the straight edges. Remember, you will have to lower the cutter to compensate for the height of the curved jig. This will ensure that the raised edges are all equal in width.

Assembling Doors

After you have cut the frames and panels, the doors can be sanded and assembled. Test the fit of the parts before you apply glue to any parts. In some cases, you may decide to stain the panels before you assemble the doors. This practice ensures that the entire panel is colored, and eliminates uncolored edges if the panel shrinks.

Fit the panel into the opening, but do not glue it. After the glue cures, pin the panel at the top and bottom on the centerline. This allows the door to expand and contract off the centerline. When the door panel is pinned, it cannot shift to one side or the other. When the panel shifts completely to one side, it could produce an opening along the door frame.

Doors with glass panels use a rabbet-style cope-and-stick joint. This technique is shown in Illus. 7-109–7-115, where it is being used on a house door. In this case, a separate moulding is needed to hold the glass in position. This moulding is mitred and nailed or screwed to the door.

Stain the insides of the rabbets and mouldings before installing the glass. This practice

Illus. 7-108. When all the curves have been raised, remove the jig and raise the straight ends and edges.

Illus. 7-109. Doors with glass panels use a rabbet-type cope-and-stick joint. The rabbet will accommodate wood or glass panels.

Illus. 7-110. Mouldings are cut from shaped edges. They are used to fit glass or wood into the openings.

Illus. 7-111. The mouldings are mitred and nailed in place. They make replacement of the wood or glass much easier.

ensures that the rabbet will be the correct color when one looks through the glass. This is also important when mirrors are installed in a door. When one looks into the mirror, the back of the rabbet is visible.

If you are nailing the mouldings in position, protect the glass with a piece of cardboard or plywood. Cardboard or plywood will deflect stray hammer blows.

Wood panels can also be installed with rabbet-style cope-and-stick joints. In this case, do not fit the panel in the opening. Pin this panel at the top and bottom along the centerline.

You can use either cope or stick cutters to make house doors. When large doors are made, the panels are usually raised on both sides (Illus. 7-116–7-128). This requires an extra shaping operation.

House doors are quite large and heavy. The joinery is usually reinforced with dowels or lag bolts. Lag bolts are driven from the outside edge of the stiles into the rails. The stiles are counterbored to accommodate the lag bolts. They are plugged after the door is assembled.

Doors are the chief product in some shops. In these shops, four or six shapers may be set up for each specialized job in cabinet or house door fabrication (Illus. 7-129 and 7-130). This eliminates setup time and ensures uniformity and safety.

Illus. 7-112. Cut the cope joint on both ends of the rails. Use a rolling table to control the workpiece.

Illus. 7-113. Power feeders are used to shape the stick joint on rails and stiles.

Illus. 7-114. Mullions (the parts separating glass or wood panels) are shaped with a power feeder.

Illus. 7-115. Attach a coped scrap to the mitre gauge. It positions the mullion for coping. The rolling table guides the work.

Illus. 7-116. A double cope-and-stick joint is often used on wooden raised-panel doors.

Illus. 7-117. A double cope is shaped on the end of this rail. The shaper must be powerful to make this cut in a single pass. A rolling table is used to control the work.

Illus. 7-118. The Panel-crafter is being used to shape the double cope on this rail. The work must be (and is) controlled absolutely for this operation.

Illus. 7-119. The double stick cutters are being positioned relative to the cope joint.

Illus. 7-120. This mullion is being shaped with the help of a power feeder. This is the only safe way to shape this mullion.

Illus. 7-121. A coped scrap is attached to the mitre gauge. This holds the mullion securely for coping. It also acts as a backing board.

Illus. 7-122. The double-raised panel is shaped in one pass. This requires a shaper with 5 horsepower, a one-inch arbor, and that you have absolute control of the workpiece.

Illus. 7-123. Use the Panelcrafter or a rolling table to raise the ends of this panel. For crowned or curved panels, use the Panelcrafter.

Illus. 7-124. After shaping the end grain, lower the spindle.

Illus. 7-125. Adjust the height of the spindle so that it is relative to the work as it sits on the table.

Illus. 7-126. Use the power feeder to shape the edges of the panel.

Illus. 7-127. Set the height of the feeder carefully. If it is too low, it will cause the panel to lift during edge shaping.

Illus. 7-128. The power feeder offers a safe and chatter-free operation.

Illus. 7-129. This four-shaper setup can be used to rapidly make cabinet doors. (Photo courtesy of L. A. Weaver)

Illus. 7-130. This six-shaper setup is more mechanized. It can be used to make house or cabinet doors. (Photo courtesy of L. A. Weaver)

Shaping Drawer Sides

You can make drawer parts on the shaper using straight cutters and roundover cutters. These parts can be joined using specialty drawer-joint cutters.

Drawer stock is typically ½ inch thick. Its width varies according to the size of the drawer.

To transform the stock into drawer parts, cut a groove or rabbet along its bottom edge (Illus. 7-131). This is to accommodate the drawer bottom. A straight cutter is used for this operation. You can either power-feed or hand-feed the piece. In production work, it is better to power-feed it. Use featherboards or other suitable hold-downs for hand-feeding operations.

Replace the split fences with a straight fence (Illus. 7-132). Secure the fence assembly to the shaper with the wooden fence clear of the cut-

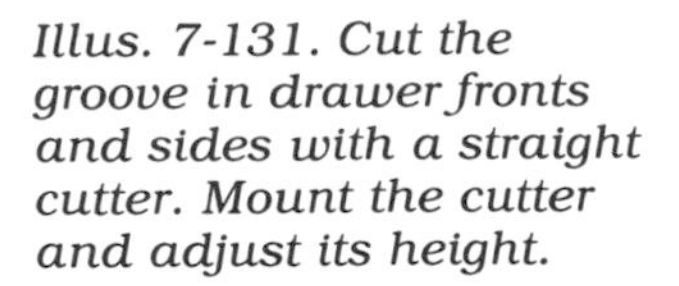

Illus. 7-131. Cut the groove in drawer fronts and sides with a straight cutter. Mount the cutter and adjust its height.

Illus. 7-132. Replace the split fences with a single board after you have mounted the straight cutter. Lock the fence assembly securely to the table and turn the shaper on.

Illus. 7-133. Back the fence into the moving cutter. This forms a slot for the cutter

Illus. 7-134. Set the fence according to the desired depth of the drawer groove. Lock the fence securely after making the setup.

Illus. 7-135. Clamp a featherboard to the shaper table. The featherboard will hold stock against the fence while you are cutting the groove.

Illus. 7-136. The distance between the fence and featherboard should be slightly less than the thickness of the workpiece. Note that the featherboard also acts as a barrier between the operator and the cutter.

Illus. 7-137. The feathers deflect slightly as the work is guided into the cutter.

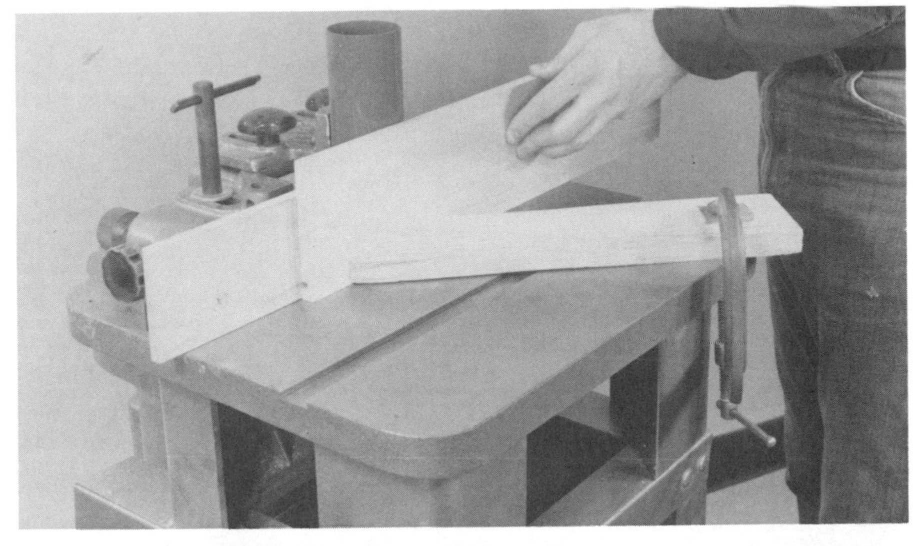

Illus. 7-138. This drawer side is being grooved in a single pass. Most shapers will make this cut in a single pass.

Illus. 7-139. This plywood drawer front is also being grooved with the same setup. The featherboard has to be moved due to the increased thickness of the drawer front.

Illus. 7-140. This solid drawer front uses the same setup. Solid stock can be fed into the cutter at a faster speed because it is less likely to tear out.

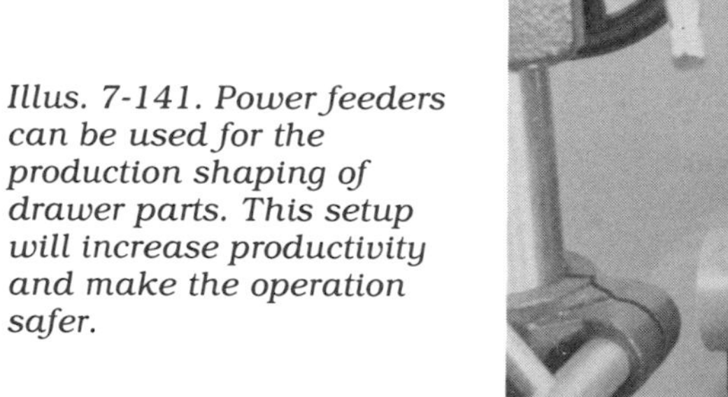

Illus. 7-141. Power feeders can be used for the production shaping of drawer parts. This setup will increase productivity and make the operation safer.

Illus. 7-142. A bead-cutting router bit is mounted in this shaper. Its radius is ideal for drawer sides.

ter. Turn the shaper on and slowly back the wooden fence into the cutter to form a slot (Illus. 7-133). Then readjust the fence to the desired depth of cut (Illus. 7-134). Set up a featherboard to hold the work against the fence, and clamp it securely (Illus. 7-135). The featherboard is clear of the cutter (Illus. 7-136) and will deflect slightly as the stock is fed into the cutter (Illus. 7-137). Once you have made the setup, turn the shaper on and proceed to shape (Illus. 7-138). The featherboard may require adjustment for other thicknesses (Illus. 7-139). Do this with the power off. Feed speed will vary according to stock species (Illus. 7-140) and the equipment available (Illus. 7-141).

On many drawers, the upper edge is radiused. The radius breaks the sharp corners and gives the drawer a more finished appearance. This can be done with a shaper (Illus. 7-142–7-149). Mount a bead cutter and adjust it in accordance with the thickness of the stock. You can hand-feed or power-feed the stock.

In some cases, the first two inches of the drawer side is left square. The remainder of the edge is radiused. This operation requires a stop block. Clamp the block to the fence at the desired position. Butt the end of the drawer side against the stop, and pivot it into the cutter. If you have already cut the groove for the drawer bottom, be sure to pair the drawer sides. Shape one piece with the groove up; shape the other piece with the groove down.

Leave the square end on the drawer. This way, the drawer side and front will join squarely. In some cases, the joint does not look as professional when the drawer side rounds over against a square shoulder.

Illus. 7-143. Mount a wooden spacer block onto the table. The spindle cannot be lowered far enough to position the bit even with the table.

Illus. 7-144. Adjust the bit so that it's relative to the drawer side.

Illus. 7-145. Adjust the fences according to the amount of stock that must be removed.

Illus. 7-146. Make a test cut on the setup piece. Make any needed adjustments before proceeding.

Illus. 7-147. Adjust the guard. Then proceed to shape all the drawer sides.

Illus. 7-148. The finished drawer side is ready for the joinery.

Illus. 7-149. This production setup can be used to radius the top of the drawer stock. Power feeders can reduce labor and boost productivity.

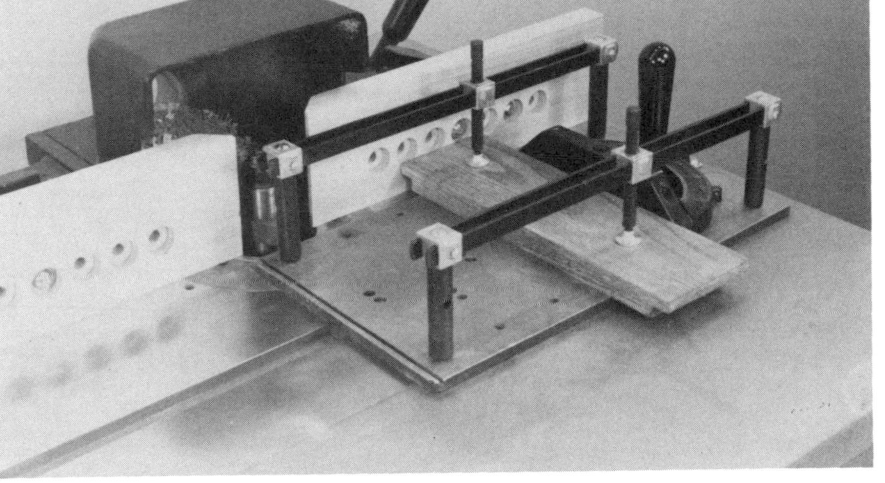

Illus. 7-150. The sliding shaper jig can be used to rabbet drawer fronts. Add a backing board to reduce tearout.

Shaping Drawer Joints

The simplest drawer joint cut on the shaper is the rabbet joint. A rabbet is cut on both ends of the drawer front. The drawer sides are glued to the rabbet. A straight cutter is used to cut the rabbet (Illus. 7-150–7-155).

Some shaper cutters are designed to cut locking drawer joints. The most common cutter is the glue-joint cutter (Illus. 7-156–7-173). Two other common drawer corner joints are the double-lock mitre (Illus. 7-174–7-184) and the drawer lock corner (Illus. 7-185–7-192). Both joints can be used to quickly and strongly join drawer sides.

To use the cutters designed for these joints, plane the workpiece to the thickness desired. Cutters of this type accommodate stock that ranges in thickness from ½ to 1⅛ inches. Use the mitre gauge and straight fences. Special jigs and fixtures can be made for production setups.

These cutters are designed to cut the drawer front and drawer sides with a single cutter setup. Cut drawer fronts horizontally, and the sides vertically. To set up the cutter, use a sample board the same thickness as the workpiece. Place the board horizontally on the shaper table and against the straight fence, and move the fence in or out to align the top of the cutting edge with the top edge of the sample board. Stand the sample board vertically and against the straight fence to adjust the cutter vertically with the spindle height adjustment. Raise or lower the cutter and align the outer cutting edge with the top edge of the sample board. It is a good idea to mark the top surface of the sample board, since it is continually being used to adjust the cutter.

Provide extra width to all workpieces being shaped so that you can joint them afterwards. You can minimize tearout by using a backing board while shaping the end grain.

Use a power-feed attachment to minimize dangerous hand feeding through cutters and prevent chatter. Adjust the feed unit to accurately hold the workpiece down on the table and against the fence in both the horizontal and vertical cuts.

It is important that the infeed and outfeed fences be as close to the cutter as possible. You can cope the fences to fit close to the cutter after you have made adjustments on the sample board. Trace a profile of the cutter onto the fence

Illus. 7-151. This jig is mounted to the mitre gauge. A backing board, as indicated by the pencil, is included to reduce tearout.

Illus. 7-152. Position the drawer front next to the jig with the groove down. The clamps hold the work securely to the jig.

face and cut it using a band saw. Allow clearance for cutter adjustments.

A special high fence can also be used to control stock shaped in the vertical position. The fence clamps to the regular fences. It has a hole in it for the cutter. Make a saddle-type fixture to go over the high fence. This fixture becomes the backing board for the drawer sides.

When you have cut the double-lock, use the saddle to feed the drawer sides into the cutter. You will have much better control, and your hands will be farther from the cutter.

The double-lock mitre is typically used on two pieces of equal thickness. If the drawer front is thicker than the drawer sides, set the cutter up for the thinner stock. This means that the parts will match up when joined. The mitre will stop short of the front edge of the drawer front.

Some drawer fronts have a bead shaped along the top and/or bottom edge. This bead makes the clearance around the drawer less visible. The space looks like part of the moulding.

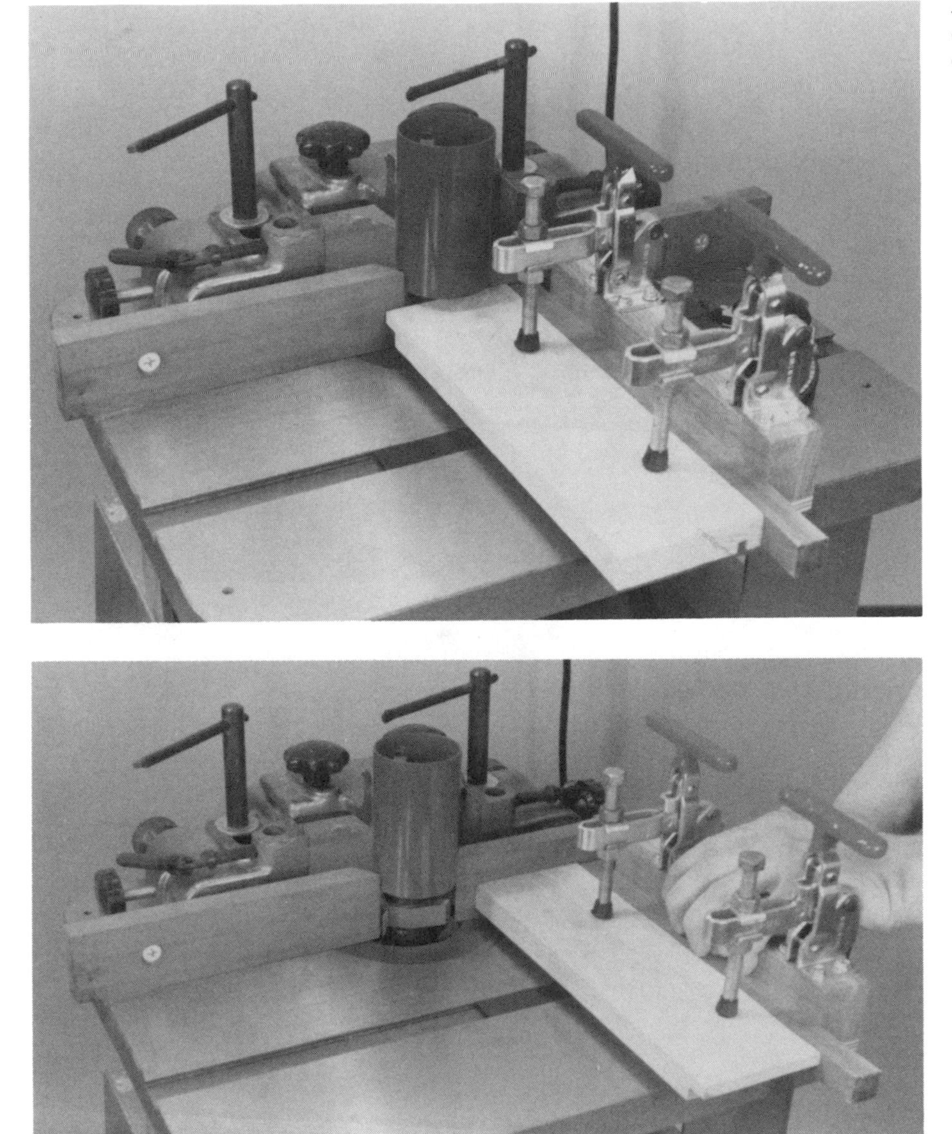

Illus. 7-153. Feed the drawer front into the cutter to form the rabbet.

Illus. 7-154. Continue shaping the opposite end of the drawer front.

Illus. 7-155. Test the fit between the drawer front and side. Make any needed adjustments.

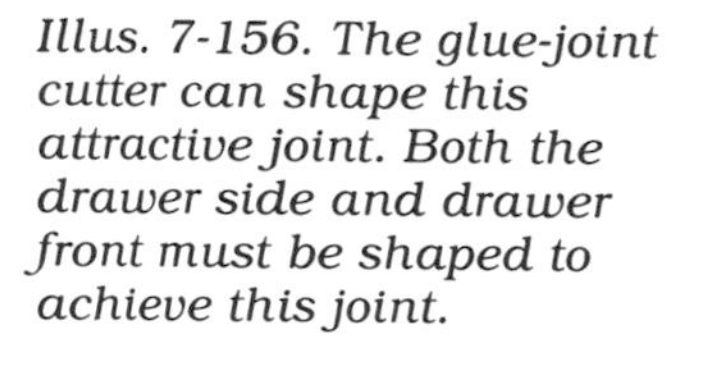

Illus. 7-156. The glue-joint cutter can shape this attractive joint. Both the drawer side and drawer front must be shaped to achieve this joint.

Illus. 7-157. Mount the glue-joint cutter on the arbor and clamp a barrier guard to the fence.

Illus. 7-158. A mitre gauge with clamps can be used to shape the end grain on the drawer front.

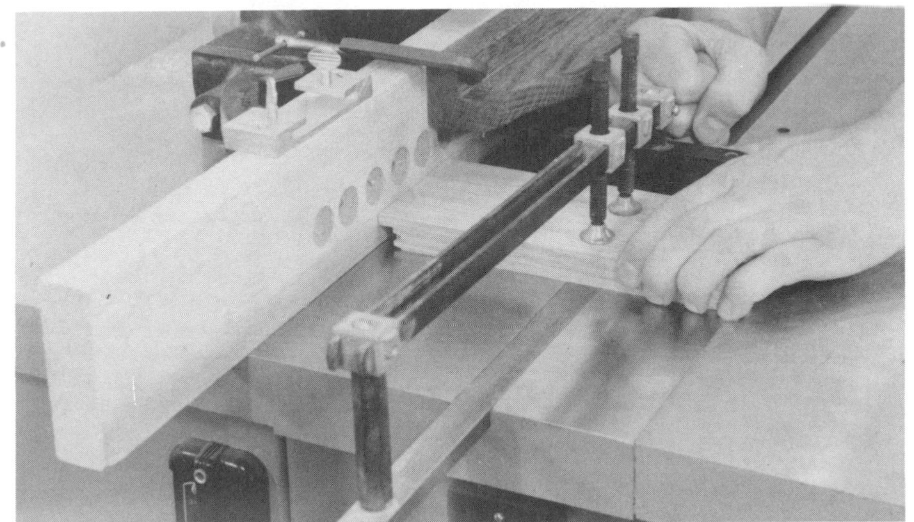

Illus. 7-159. The clamps help control the workpiece while it is being shaped.

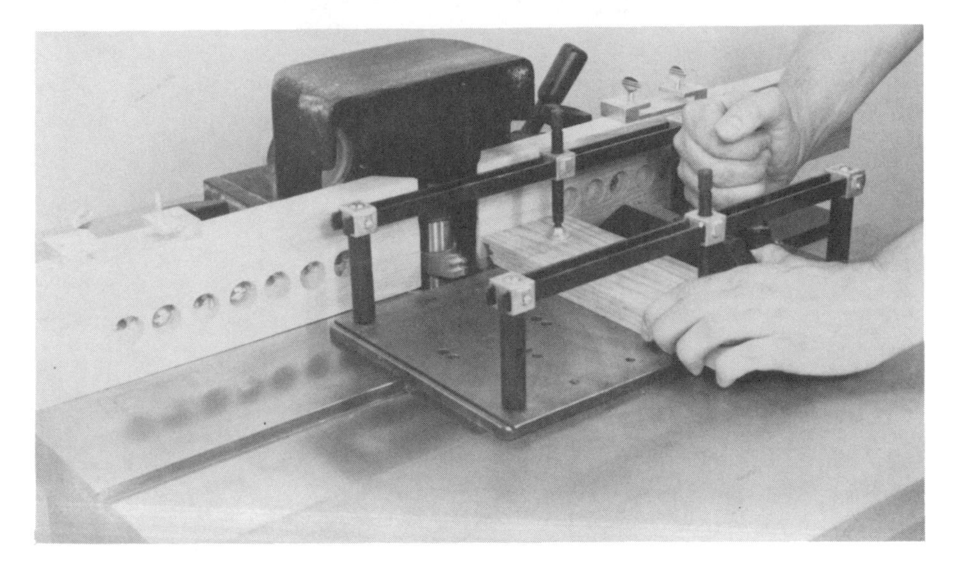

Illus. 7-160. The sliding shaper jig can also be used to shape the drawer fronts. You have to raise the cutter to make this cut.

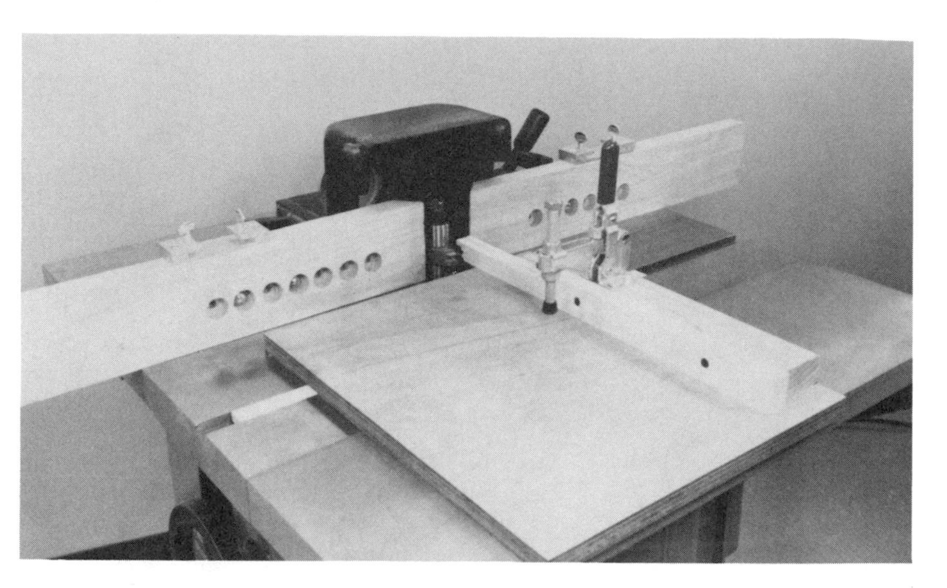

Illus. 7-161. This shop-made jig can also be used to shape the drawer front. The backing board will control tearout.

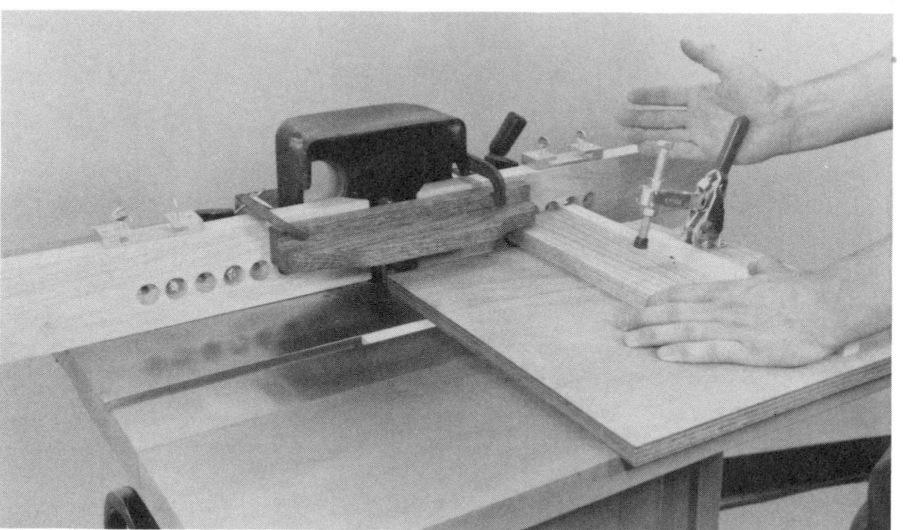

Illus. 7-162. Position the barrier guard and clamp the work to the jig.

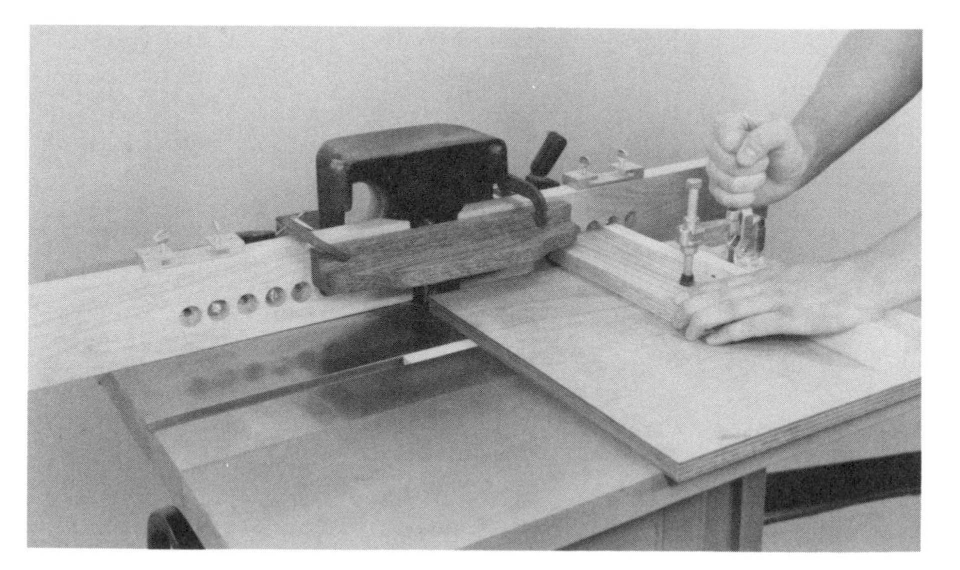

Illus. 7-163. Guide the drawer front into the cutter.

Illus. 7-164. It is important that you control the stock while shaping end grain. Make the cuts safely. Clamps help control stock effectively.

Illus. 7-165. Once you have shaped the drawer front, you must shape the drawer sides for a tight fit.

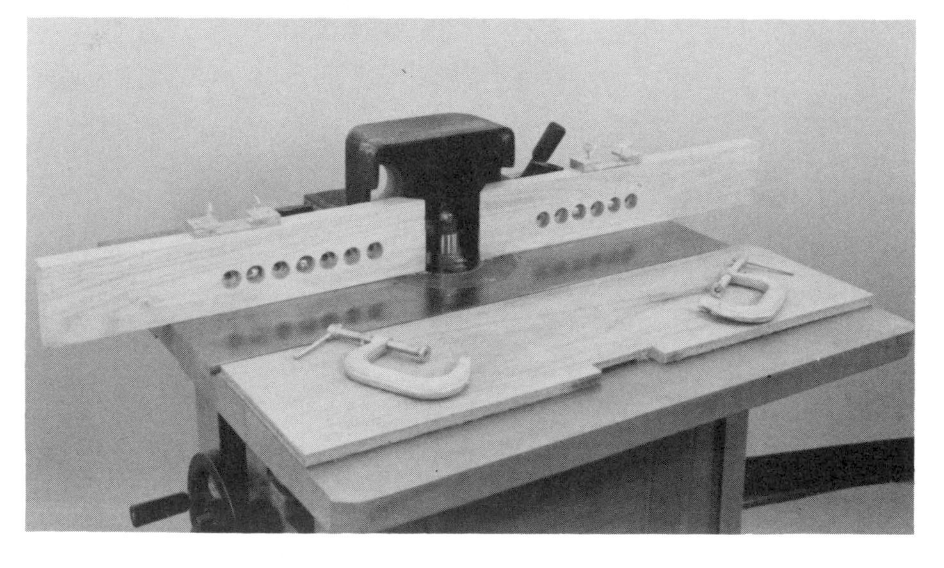

Illus. 7-166. To control the drawer sides during shaping, use an auxiliary fence.

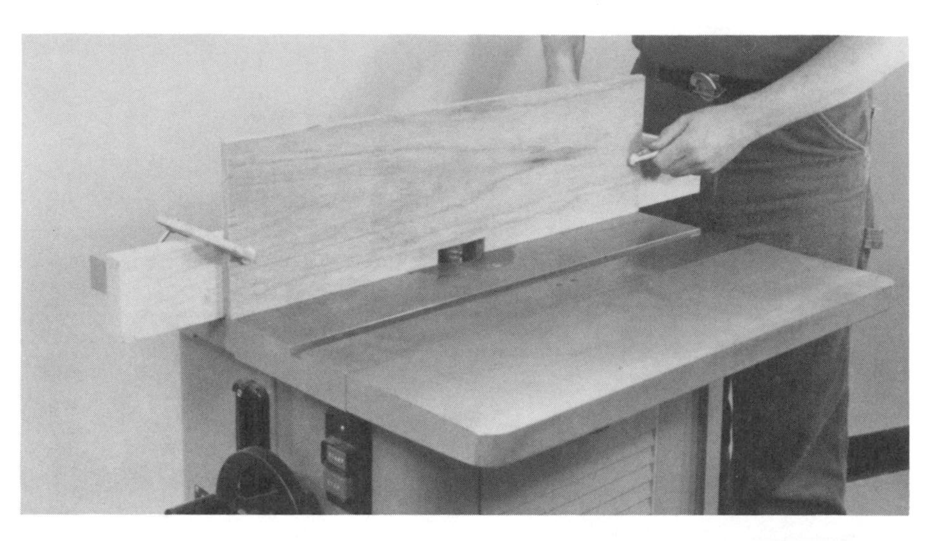

Illus. 7-167. Clamp the auxiliary fence to the shaper fence.

Illus. 7-168. Fit a push block to the auxiliary fence. The push block slides on the top of the auxiliary fence.

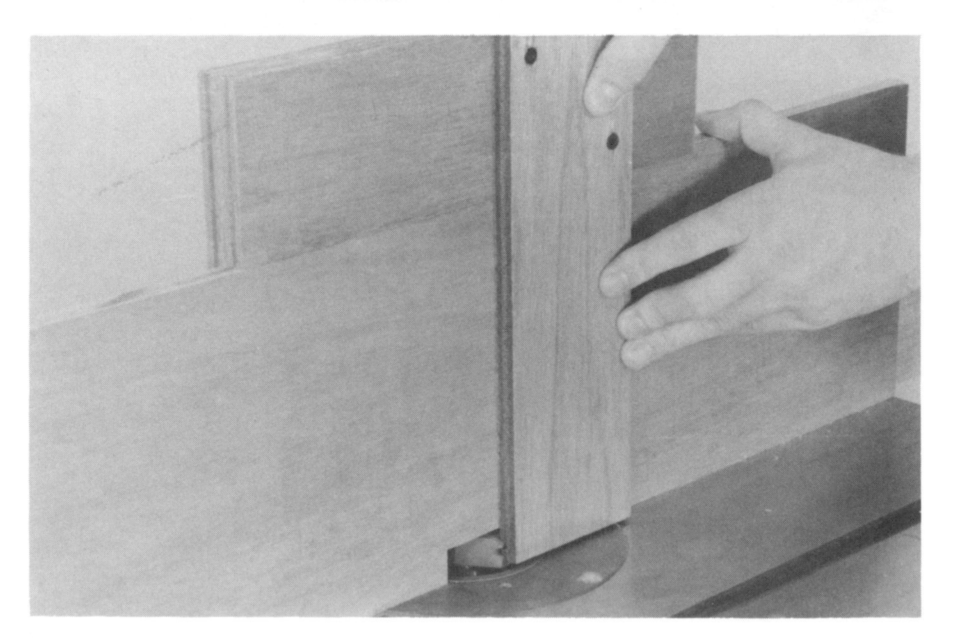

Illus. 7-169. Shape the push block before using it. It can be used to control tearout.

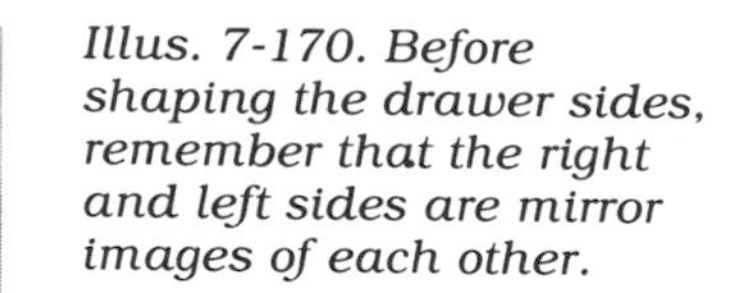

Illus. 7-170. Before shaping the drawer sides, remember that the right and left sides are mirror images of each other.

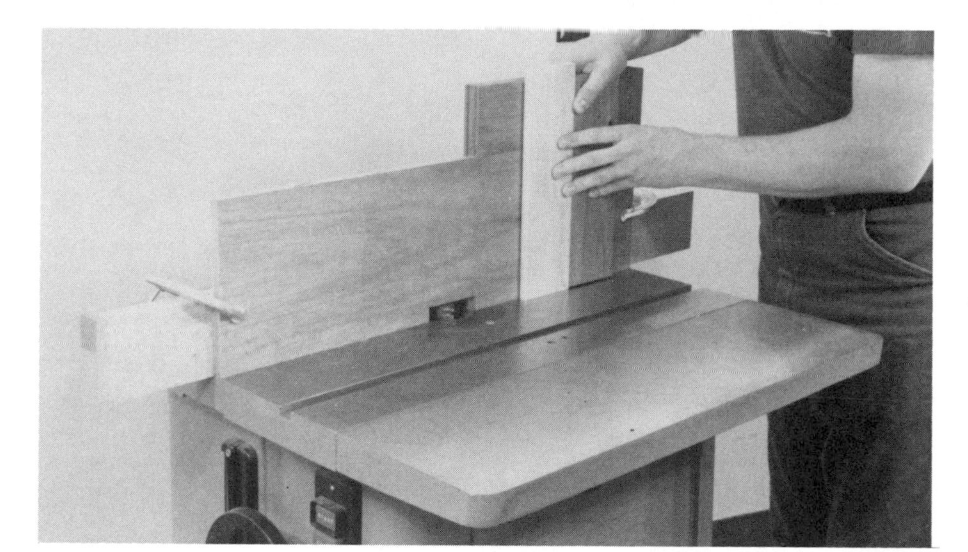

Illus. 7-171. Position the drawer side against the fence and push block.

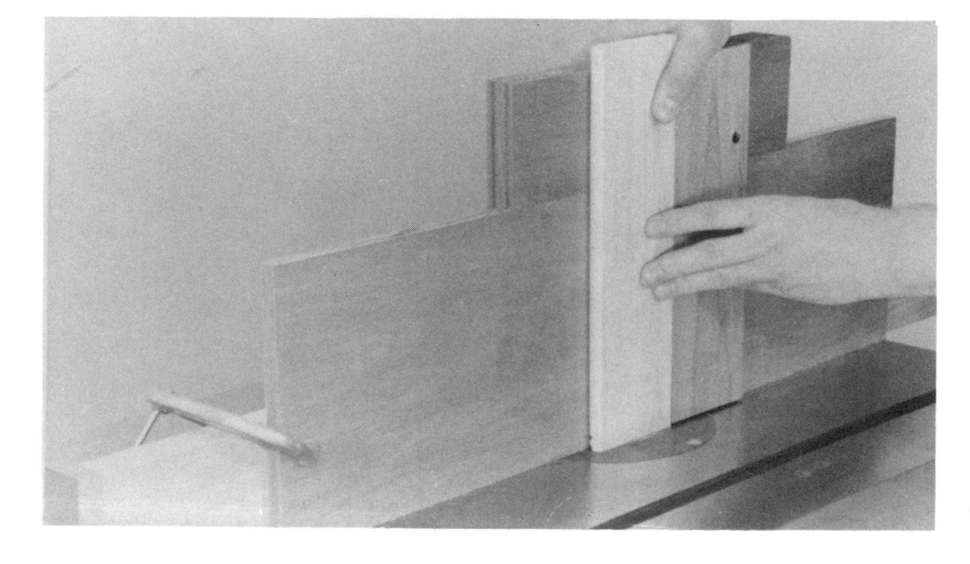

Illus. 7-172. Guide the drawer side into the moving cutter. Keep it against the fence and push block. Remember, the groove should face the fence on both drawer sides.

Illus. 7-173. Test the fit between parts, and make any necessary adjustments.

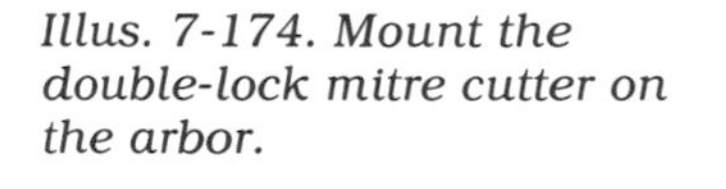

Illus. 7-174. Mount the double-lock mitre cutter on the arbor.

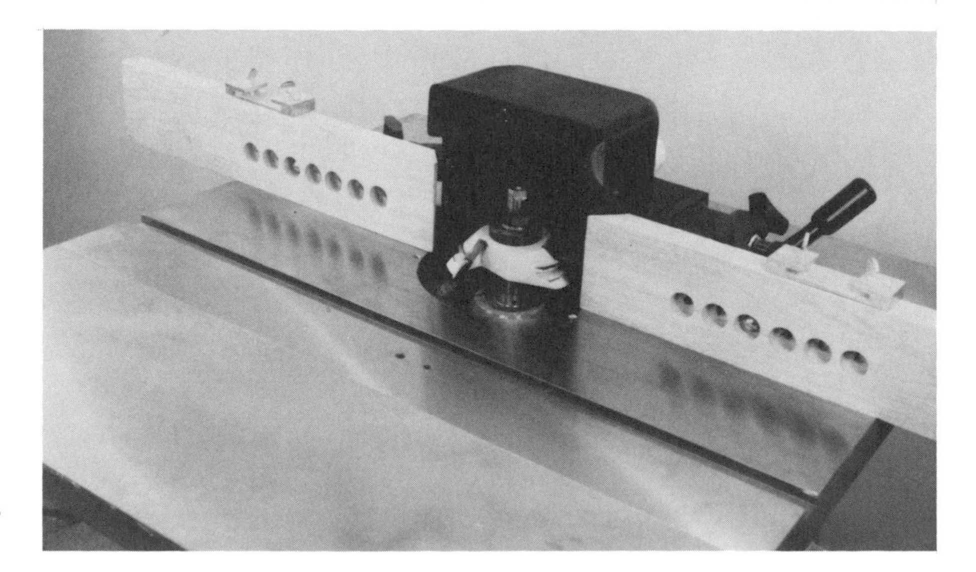

Illus. 7-175. Mount the fence assembly on the arbor. You have to replace the fence boards.

Illus. 7-176. Replace the split fences with a special coped single-piece fence.

Illus. 7-177. Use a special shop-made jig to control the stock. The jig rides along the fence.

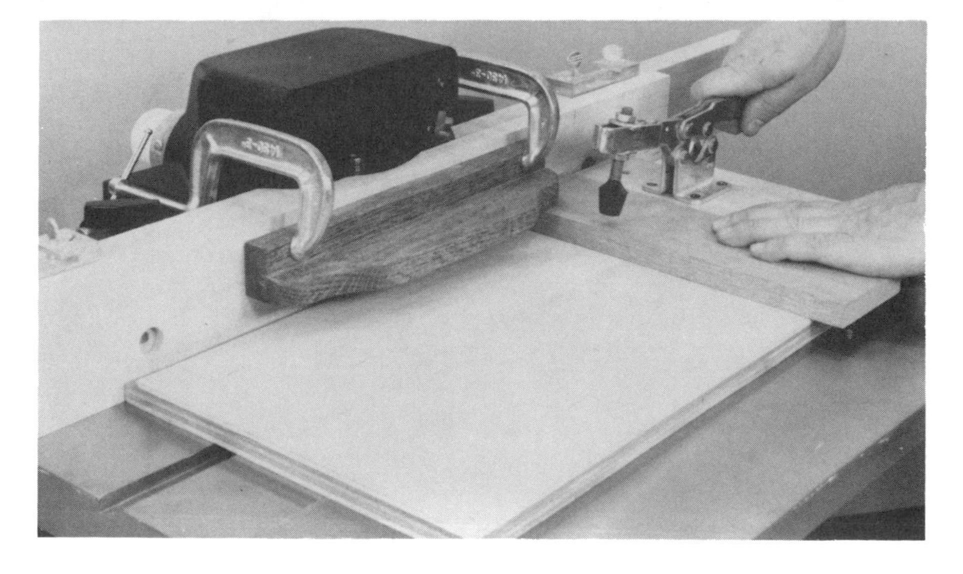

Illus. 7-178. The drawer fronts are usually done in the horizontal plane. The clamp holds the stock securely while it is being shaped.

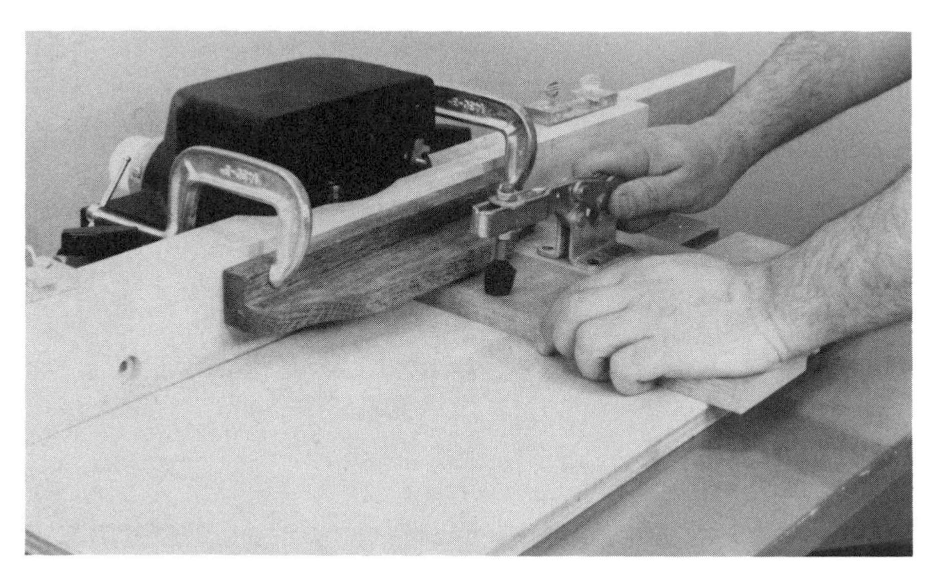

Illus. 7-179. Keep the jig against the fence as you continue shaping. The barrier guard makes contact with the cutter difficult.

Illus. 7-180. Add a special table and fence to the shaper so that you can shape the drawer sides.

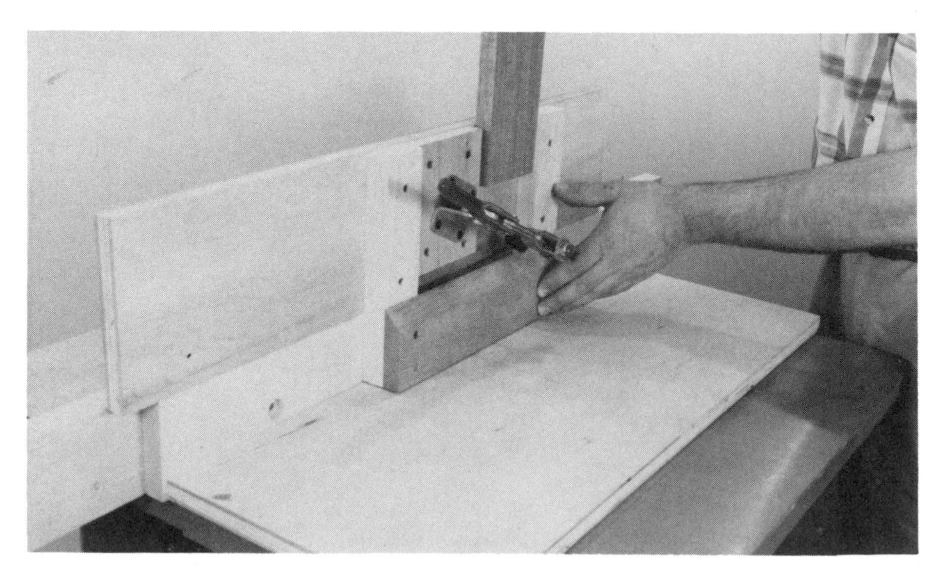

Illus. 7-181. The shop-made clamping jig rides on the fence and holds the piece perpendicular to the cutterhead.

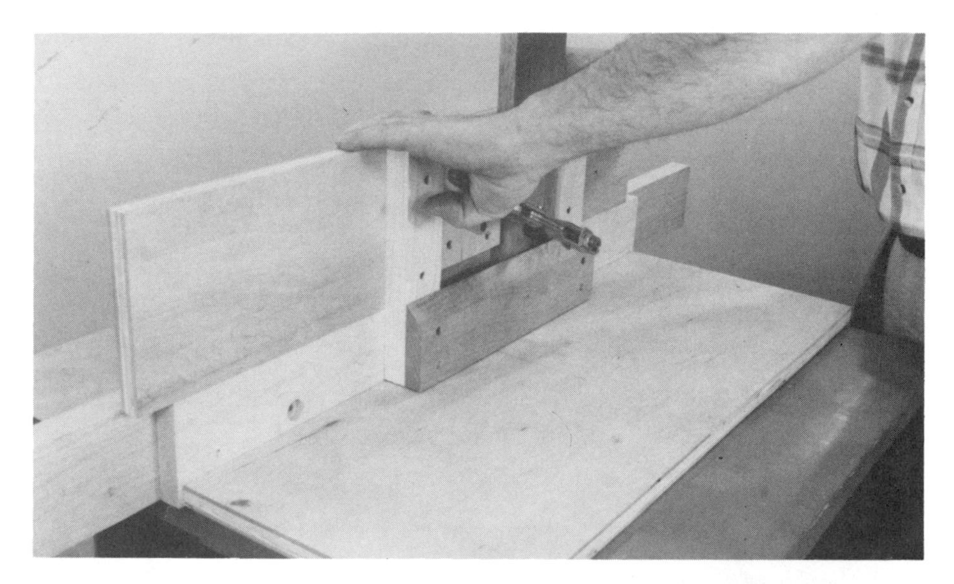

Illus. 7-182. Lock the clamp against the workpiece. Make sure that the workpiece is positioned correctly.

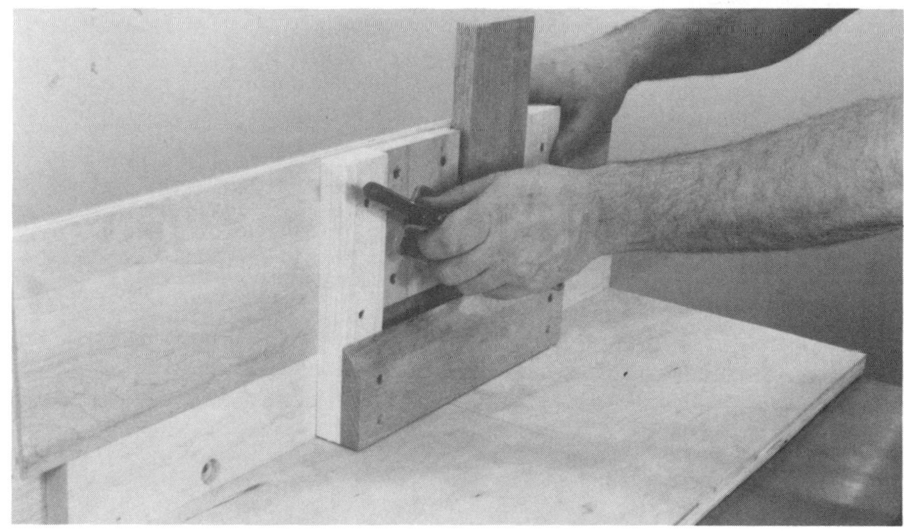

Illus. 7-183. Guide the jig along the fence top, keeping the workpiece against the face of the fence.

Illus. 7-184. Reverse the workpiece to shape the opposite end. If only the front end is shaped, be sure that you shape a right-hand and left-hand drawer side.

Illus. 7-185. This drawer lock corner joint can be made with a single cutter. The drawer front is shaped in the horizontal plane, and the drawer side is shaped in the vertical plane.

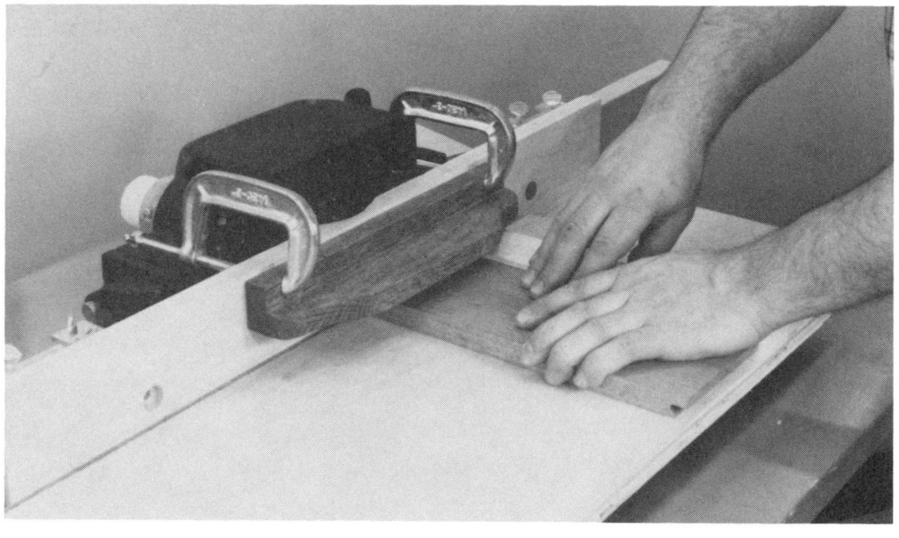

Illus. 7-186. Use a shop-made sliding jig to shape the end of the drawer front.

Illus. 7-187. This joint requires a light cut, and very little cutter extends through the coped fence. This means that hand-clamping pressure is all that is required for this operation.

Illus. 7-188. A shop-made jig is used to shape the drawer sides. It slides on the top of the shaper fence.

Illus. 7-189. Slip the drawer side into the jig. The groove for the drawer bottom faces the arbor.

Illus. 7-190. Lock the drawer side securely in position. Make sure that it is positioned correctly.

Illus. 7-191. Push the jig across the fence, keeping the work against the face of the fence.

Illus. 7-192. Test the fit between the parts; make any needed adjustments.

8 SPECIALIZED OPERATIONS

During the course of writing this book, Roger Cliffe and Michael Holtz encountered several special challenges while using the shaper. These challenges and their solutions are explored here in the hope that they will prove informative and inspire you to use the shaper to its fullest capabilities.

Lock Mitre Posts

One special challenge undertaken by the authors was to make decorative posts that could actually go over softwood support posts. This job required some type of post-and-beam construction. The double-lock mitre cutter was used for this operation. After the posts were glued up, additional cuts were taken.

The double-lock mitre cutter proved to be ideal for box construction or post-and-beam construction. The boxes glue up easily, since there is a positive lock between the parts. Cut the stock to the desired width.

After mounting the cutter to the spindle, adjust its height relative to the thickness of the stock. Cope the fences and move them close to the cutter. Adjust them so that the entire edge of the board will be cut away.

Adjust the power feed (Illus. 8-5), and cut the parts. Make a test cut to be sure that the cutter is positioned correctly.

Four parts are required to make a column. Two parts can be shaped horizontally, and two pieces can be shaped vertically (Illus. 8-6 and 8-7). An alternative method would be to shape one half of all the parts horizontally, and one half of all the parts vertically.

You can shape columns if you use the correct cutters. The process takes two cuts per face. Mount a bead or half-round cutter of the desired diameter to the spindle and cut the square stock (Illus. 8-9). The thickness and width of the stock should be the same as the diameter of the column.

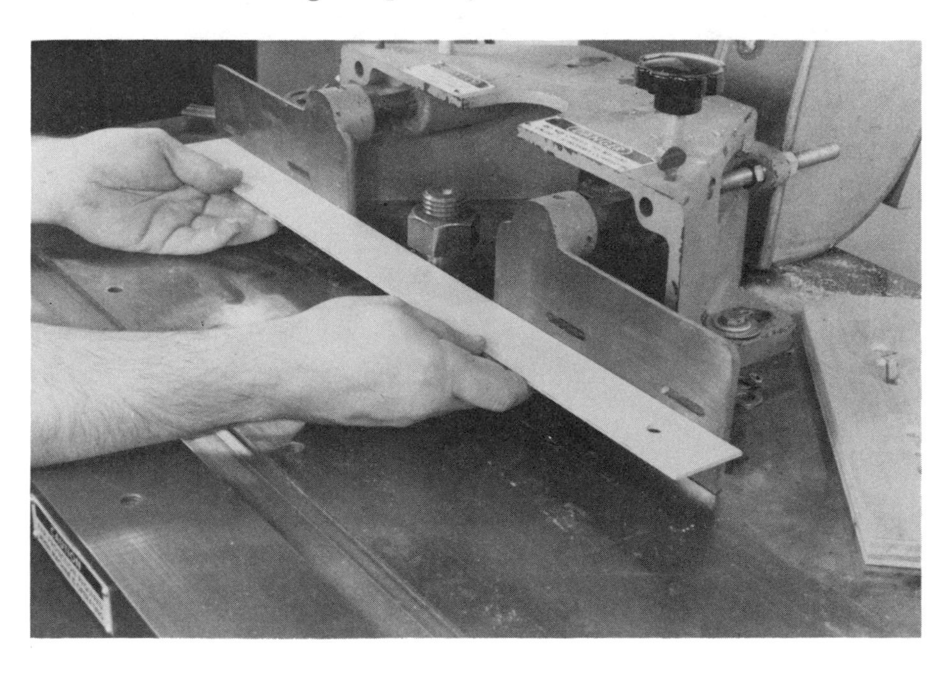

Illus. 8-1. Set the fences in a true plane to cut lock mitres.

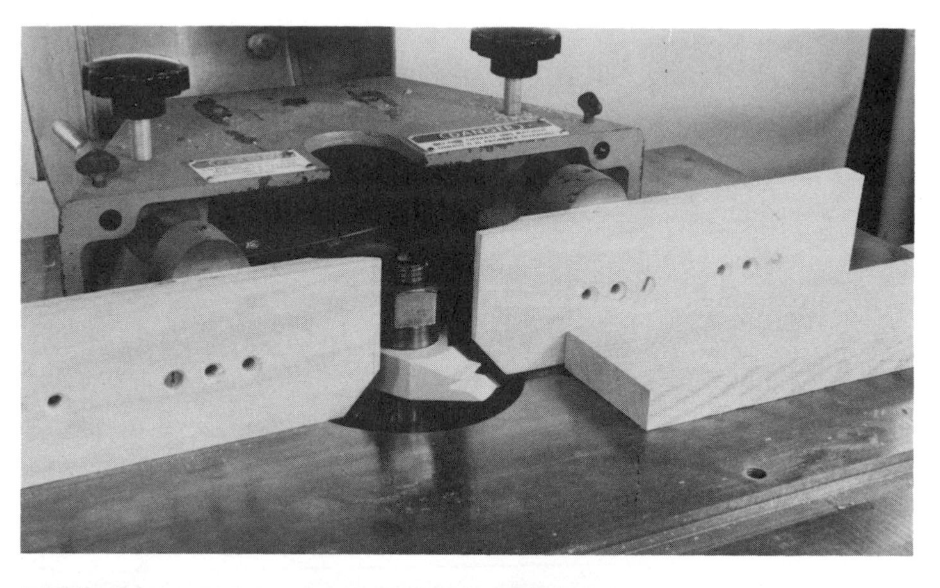

Illus. 8-2. Cope or cut the wooden fences and move them closer to the cutter. This will limit the size of their chips.

Illus. 8-3. Adjust the height of the cutter so that it is relative to the thickness of the stock.

Illus. 8-4. Move the power feed into position. Adjust its wheels for the thickness of the stock.

Illus. 8-5. Position the wheels so that they hold the stock against the fence. Be sure that the cutter clears the wheels on the power feeder.

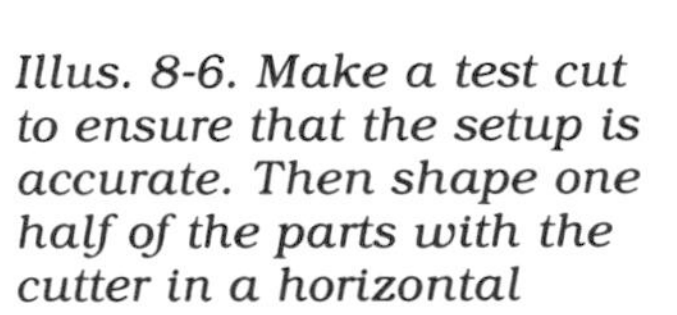

Illus. 8-6. Make a test cut to ensure that the setup is accurate. Then shape one half of the parts with the cutter in a horizontal plane.

Illus. 8-7. Shape the other half of the parts with the cutter in a vertical position.

Illus. 8-8. Position the power feeder to keep the stock down on the table and in against the fence.

Illus. 8-9. Feed the square stock into the half-round cutter.

Illus. 8-10. Cut the profile so that it's one-eighth the size of the column. Later you will make additional cuts.

Cope the fences and position them close to the cutter. Align the fences with the deepest part of the cutter. Use a power-feed unit to feed the work (Illus. 8-10).

After shaping the first edge, turn the piece end over end and shape it along the opposing edge. Continue shaping until the entire column is shaped (Illus. 8-11 and 8-12).

Next, cut the columns to the desired lengths. The support column can actually be inserted through the column for structural support.

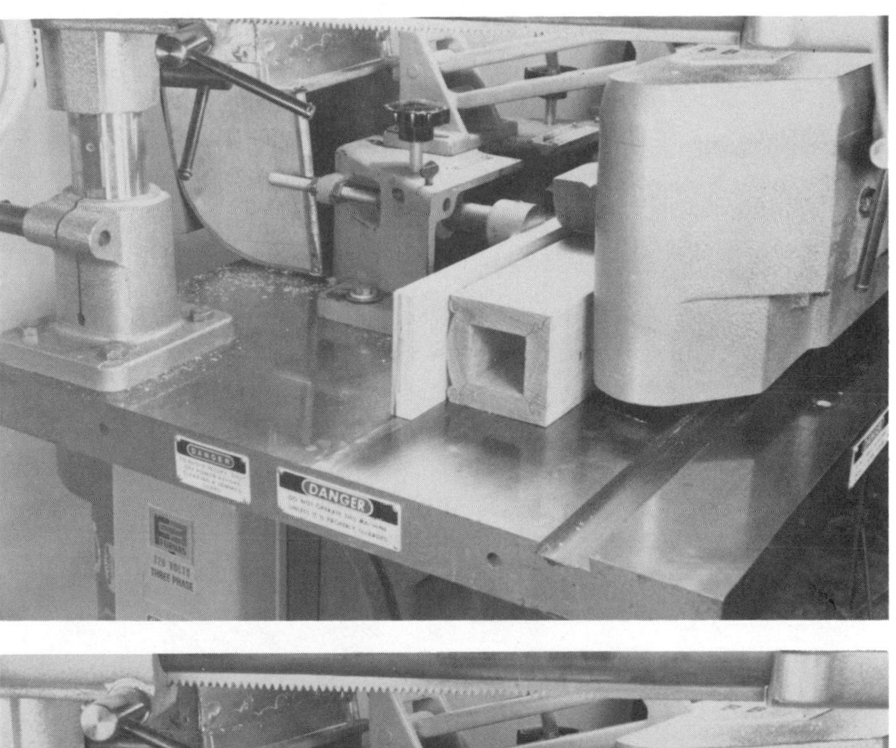

Illus. 8-11. Turn the column end over end and shape it again.

Illus. 8-12. The completed column. These columns are much faster to shape than turned columns.

Gazebo Porch Railings and Balusters

A restoration job undertaken by the authors required replacing the handrails on a gazebo porch. The rails had to be curved and shaped so that they were comfortable to the hand. In addition, budget constraints required that the balusters be shop-made and inexpensive. To keep tooling costs low, the profiles were selected from the existing collection of cutters. Following is the step by step procedures that were used to replace the handrails.

MAKING THE HANDRAILS

The straight handrails were shaped first. A raised panel bit was chosen to shape the handrail (Illus. 8-13). A crown profile was obtained when two passes were made (Illus. 8-14).

After all curved handrails and footrails were made, inside-curve and outside-curve (Illus. 8-15) fences were fabricated. The fences were cut away around the shaper cutter for clearance (Illus. 8-16).

The inside edge of the curved handrails was shaped first (Illus. 8-17). Extension rollers were used to support the end of the workpiece as the cut was completed (Illus. 8-18). When we gained experience making the cut, the power feeder was

Illus. 8-13. A raised panel cutter was used to shape the curved and straight handrails.

Illus. 8-14. One pass was made from each edge. This gave the handrail a crown profile.

Illus. 8-15. This wooden fence was custom-made to shape the inside curve on the handrails.

Illus. 8-16. The fence was cut away to accommodate the shaper cutter. This was done with a band saw.

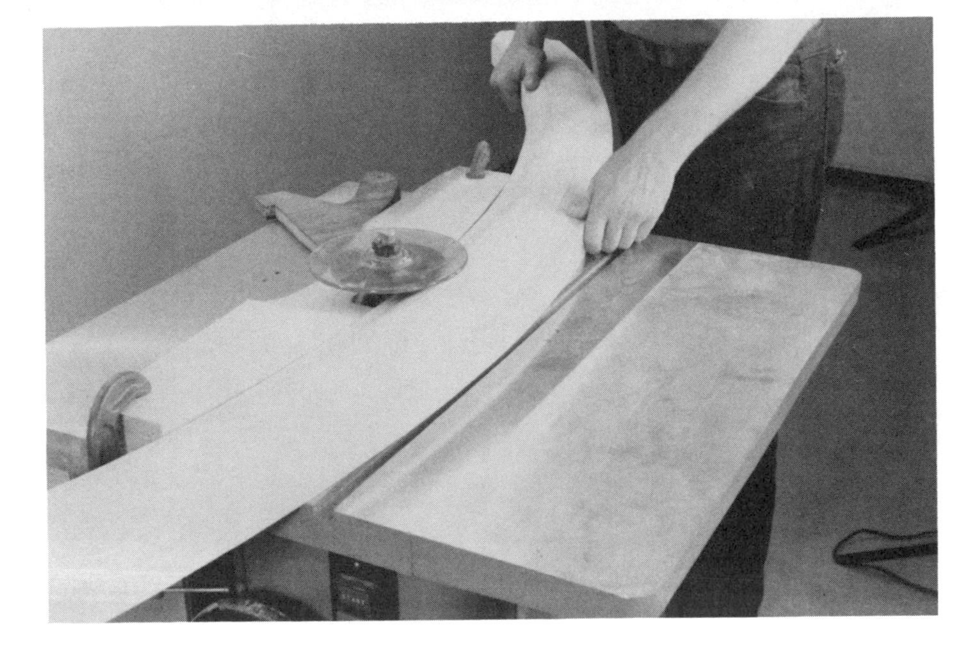

Illus. 8-17. The inside curve rides along the shop-made fence. The cutter has a chip-limiting profile to make it safer to hand-feed the stock.

Illus. 8-18. The handrails were long, so a take-off support was needed.

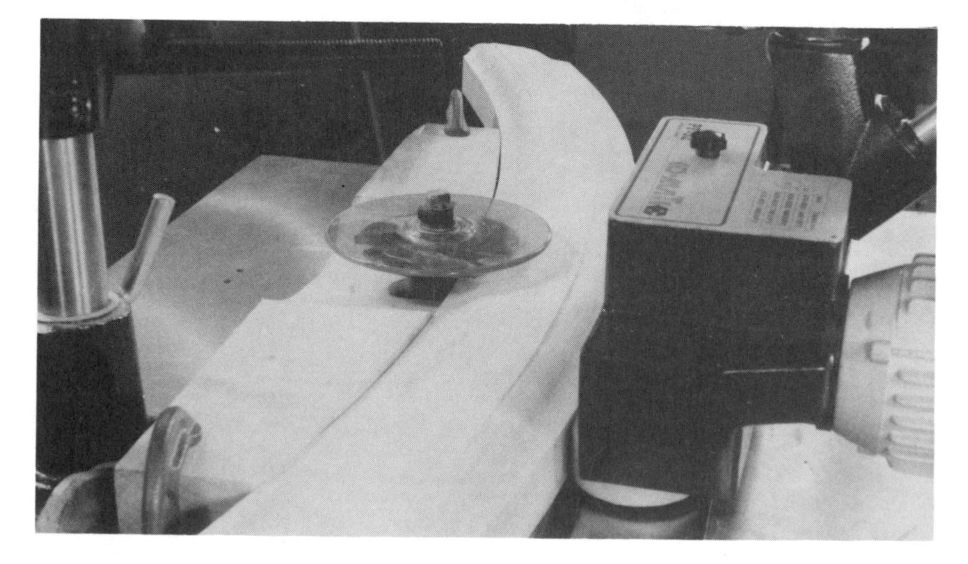

Illus. 8-19. The power feeder made the operation safer and more efficient.

Illus. 8-20. The inside curve fence was replaced with an outside curve fence. This allows the opposite side of the curved handrails to be shaped.

Illus. 8-21. A single-piece fence was used to shape the balusters. Note the cutaway area for the spindle.

added to the operation (Illus. 8-19). The wheels of the power feeder handled the curve nicely. The outside curve fence was adjusted for the opposite side of the handrail (Illus. 8-20) and it was shaped.

MAKING THE BALUSTERS

Turned balusters were expensive, so a cove cutter was selected and mounted onto the arbor. A special single-piece fence was used to replace the split fences (Illus. 8-21). The stops determined where the cove cut began and ended (Illus. 8-22 and 8-23).

Use the following procedure the writers used to shape balusters. Butt the baluster blank against the infeed stop (Illus. 8-24). Slowly push it into the cutter (Illus. 8-25). Guide the blank into the cutter towards the outfeed stop (Illus. 8-26). Once the blank touches the stop (Illus. 8-27), move it away from the cutter (Illus. 8-28).

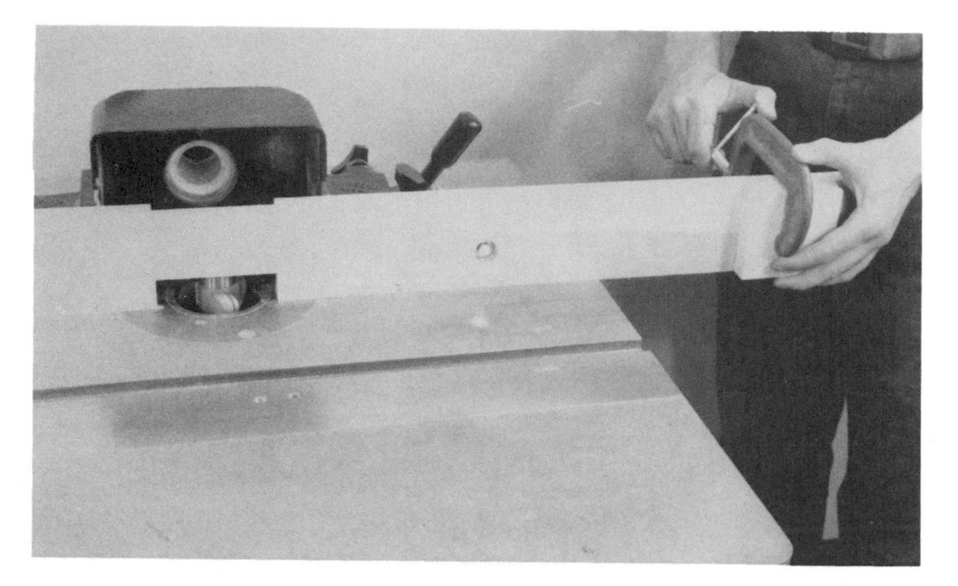

Illus. 8-22. Stops were clamped onto the fence. These stops positioned the cove cuts on the balusters.

Illus. 8-23. The fence extends beyond the ends of the shaper. This allows stops for the beginning and end of each cut.

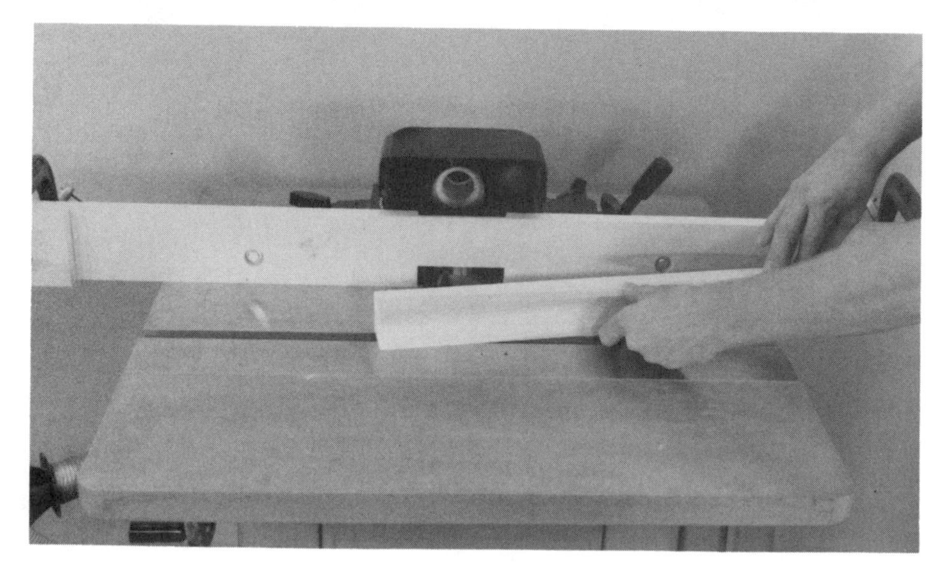

Illus. 8-24. Begin by butting the blank against the infeed stop.

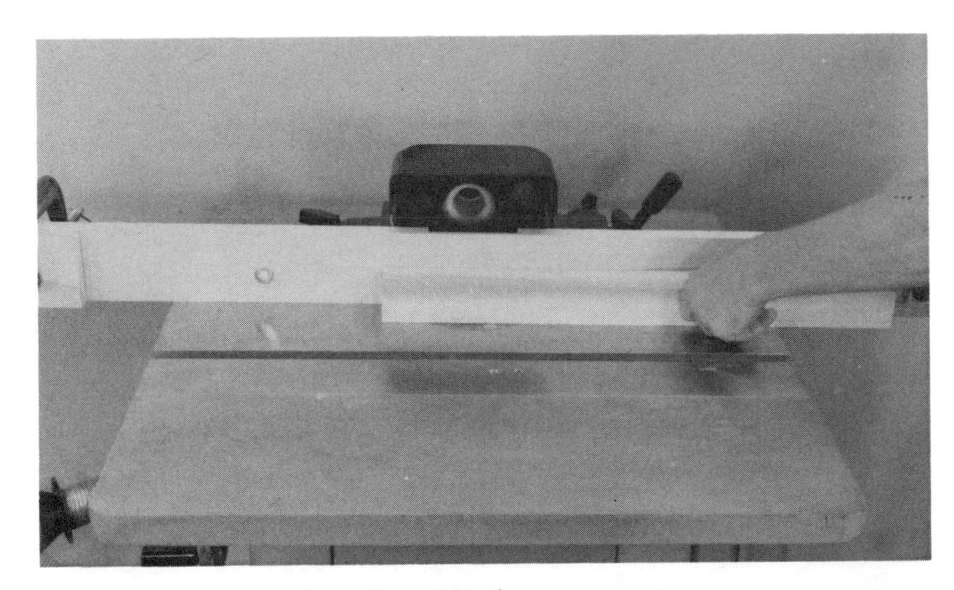

Illus. 8-25. Slowly push the blank into the cutter. Keep it against the infeed stop.

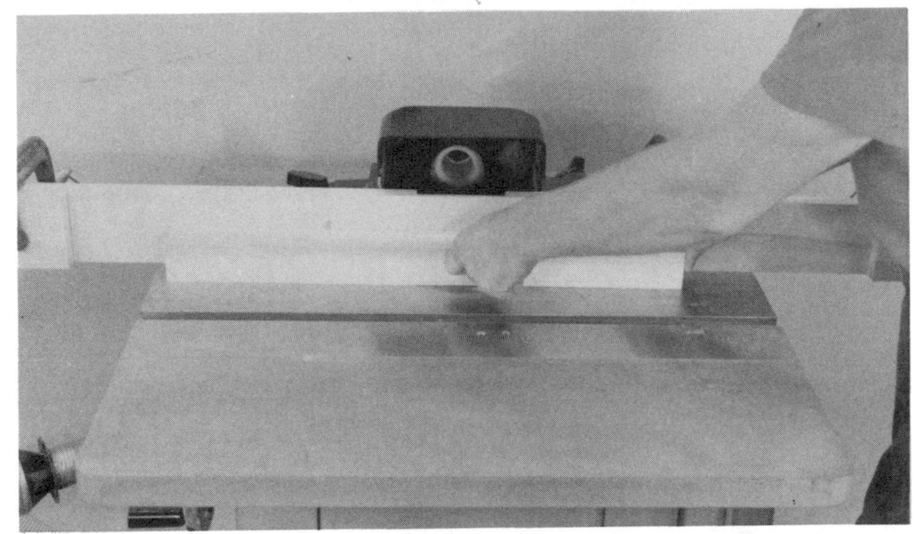

Illus. 8-26. Push the blank towards the outfeed stop, keeping it against the fence and table.

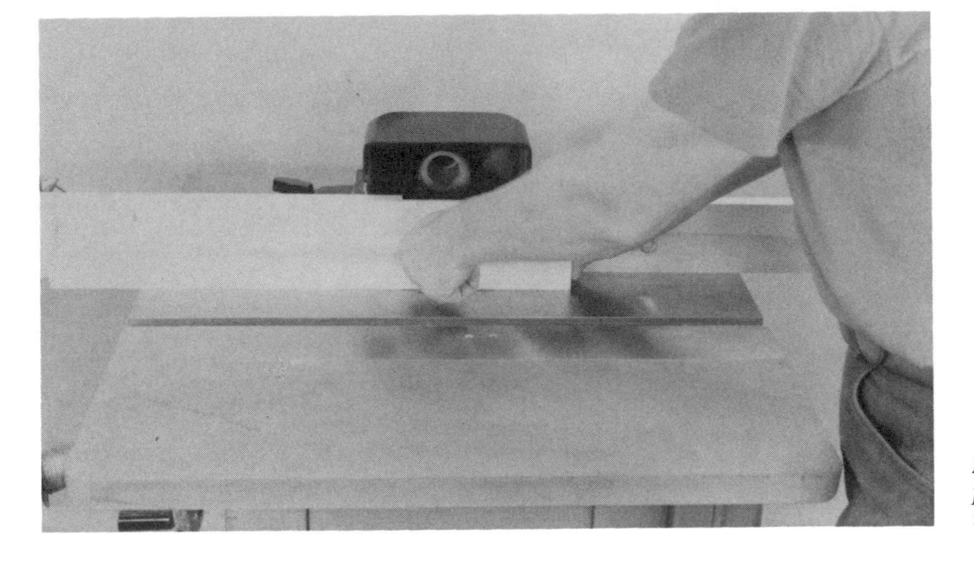

Illus. 8-27. Continue pushing the blank until it touches the outfeed stop.

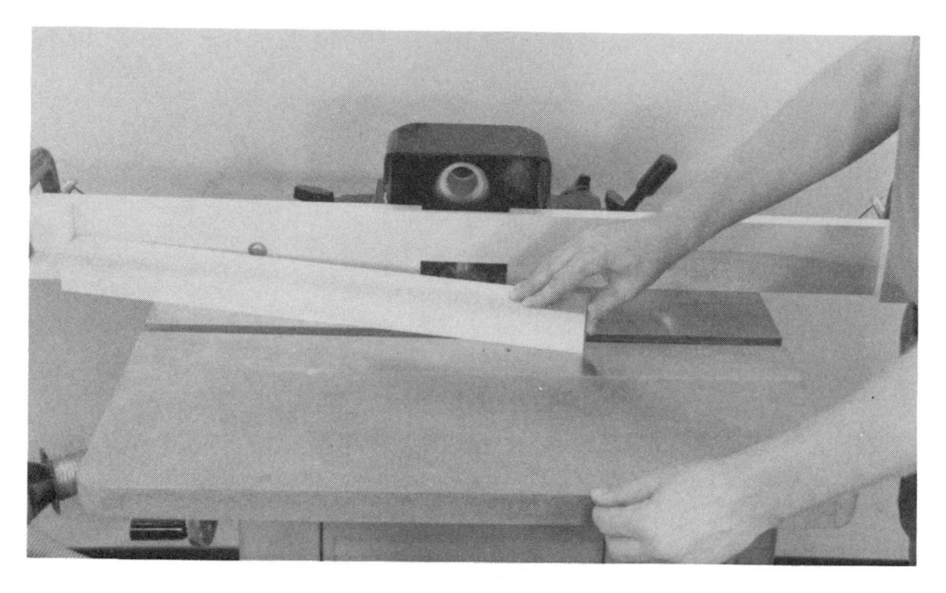

Illus. 8-28. Keep the blank against the stop and push it away from the cutter.

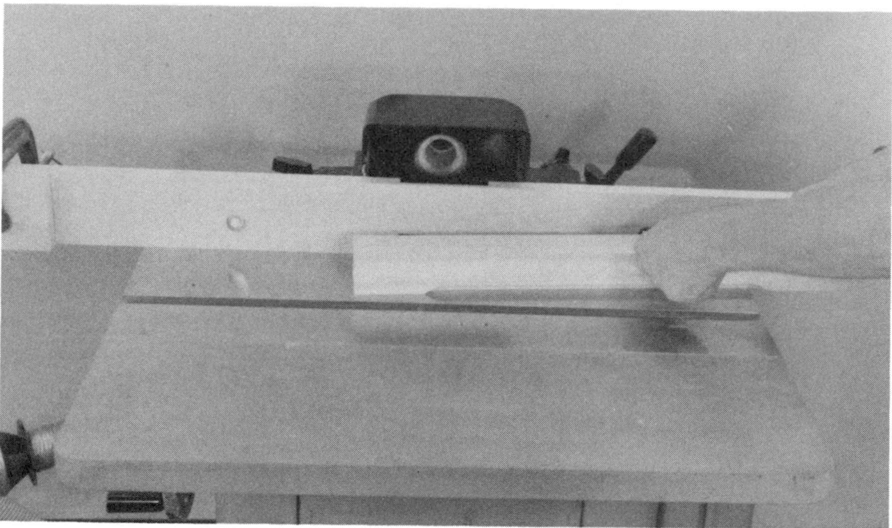

Illus. 8-29. Turn the blank 90 degrees and repeat the process. All four corners must be shaped.

Illus. 8-30. This baluster has a nice profile, and was made inexpensively.

Then turn it 90 degrees and butt it against the infeed fence (Illus. 8-29). Repeat the process on all four corners, and the baluster will take shape (Illus. 8-30).

To get the best appearance for the baluster, turn it 45 degrees on the rails. This displays more detail to anyone looking at it from the street.

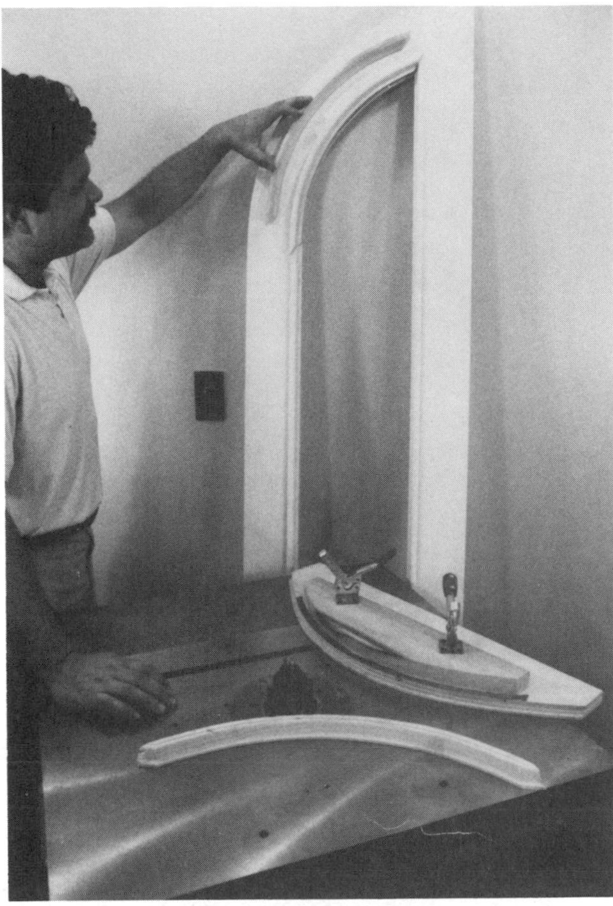

Illus. 8-31. This curved moulding presents an extra challenge when it is rabbeted.

Reversing Spindle Rotation

There are times when a reversible spindle shaper can be very useful. Consider the challenge of shaping the curved moulding on the door shown in Illus. 8-31. The grain goes the long way (Illus. 8-32). This means that the short grain will be shaped, so the rabbeted edge will be difficult to shape without a reversible spindle. That is because the grain is so short that it will tear out.

To shape the moulding, do the following: begin with a sawn blank. Shape its inside edge first (Illus. 8-33). Use a fixture to control the blank (Illus. 8-34). No starting pin is required because the fixture or template is longer than the blank (Illus. 8-35). This allows the template to contact the ball bearing before you begin shaping.

Illus. 8-32. The grain goes the long way in this frame moulding. This means that short grain will be shaped. This requires a reversible spindle shaper.

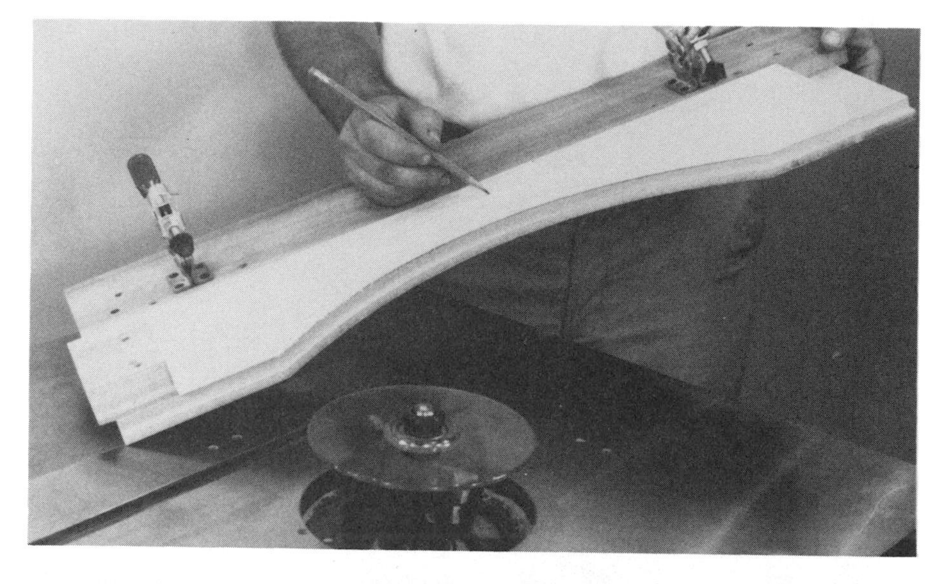

Illus. 8-33. The first step is to saw the blank to the desired inside profile.

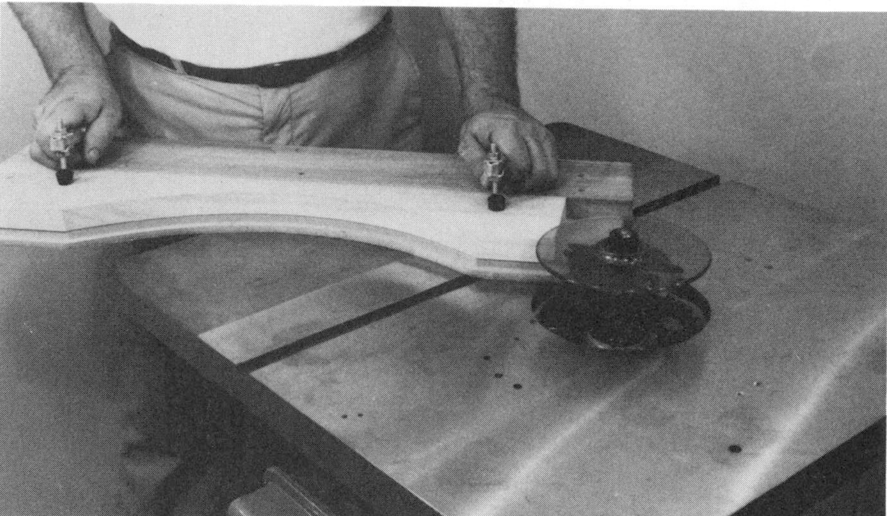

Illus. 8-34. This fixture is used to control the shape of the blank.

Illus. 8-35. No starting pin is required because the fixture is longer than the work. The fixture contacts the ball-bearing depth collar.

Illus. 8-36. Shape the entire inside profile. Note that a chip-limiting cutter is being used for hand feeding.

Illus. 8-37. A backing block on the template acts as a stop and locating pin.

Illus. 8-38. Saw and sand the moulding to the desired profile after shaping the outside edge.

Illus. 8-39. A special template has been fabricated to shape the rabbet on the short grain.

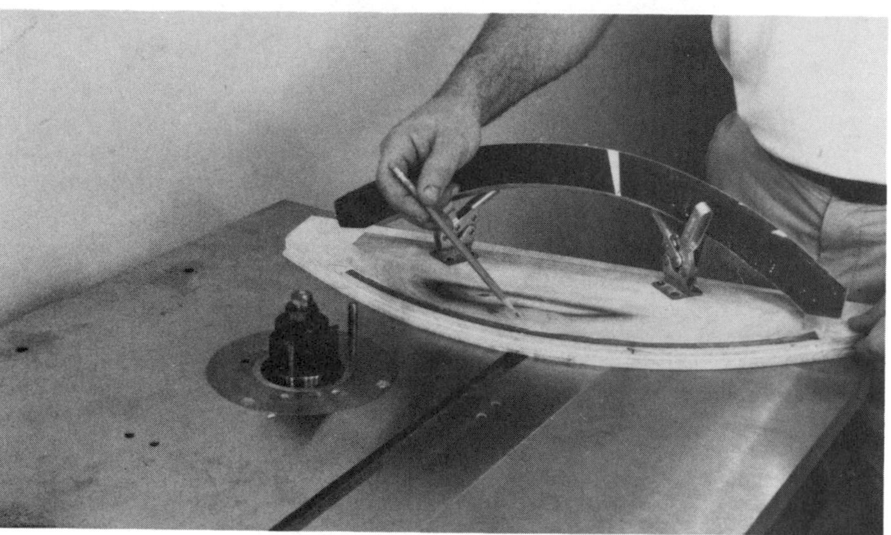

Illus. 8-40. Glue abrasives to the clamping surfaces to ensure a tight grip.

Illus. 8-41. Clamp the parts securely in the template. Note the centerline mark on the upper portion of the template. It marks the position where shaping begins.

Shape the blank completely along its inside edge (Illus. 8-36). Use a backing block to locate the work (Illus. 8-37).

After shaping the inside profile shape, saw the moulding and sand it to its final profile (Illus. 8-38). Make a special template to shape the rabbet on the short grain (Illus. 8-39). Glue abrasives to the clamping surfaces to ensure a tight grip (Illus. 8-40). Mark the jig with a centerline (Illus. 8-41).

Position a starting pin on the thrust side of the cutter. Begin the shaping on the centerline of the workpiece (Illus. 8-42). Once the template engages with the rub collar, pull it away from the starting pin (Illus. 8-43).

Form the rabbet with a straight cutter (Illus. 8-44). Comb the end grain downward to control tear out. The cut will be completed with the ball bearing riding against the template (Illus. 8-45).

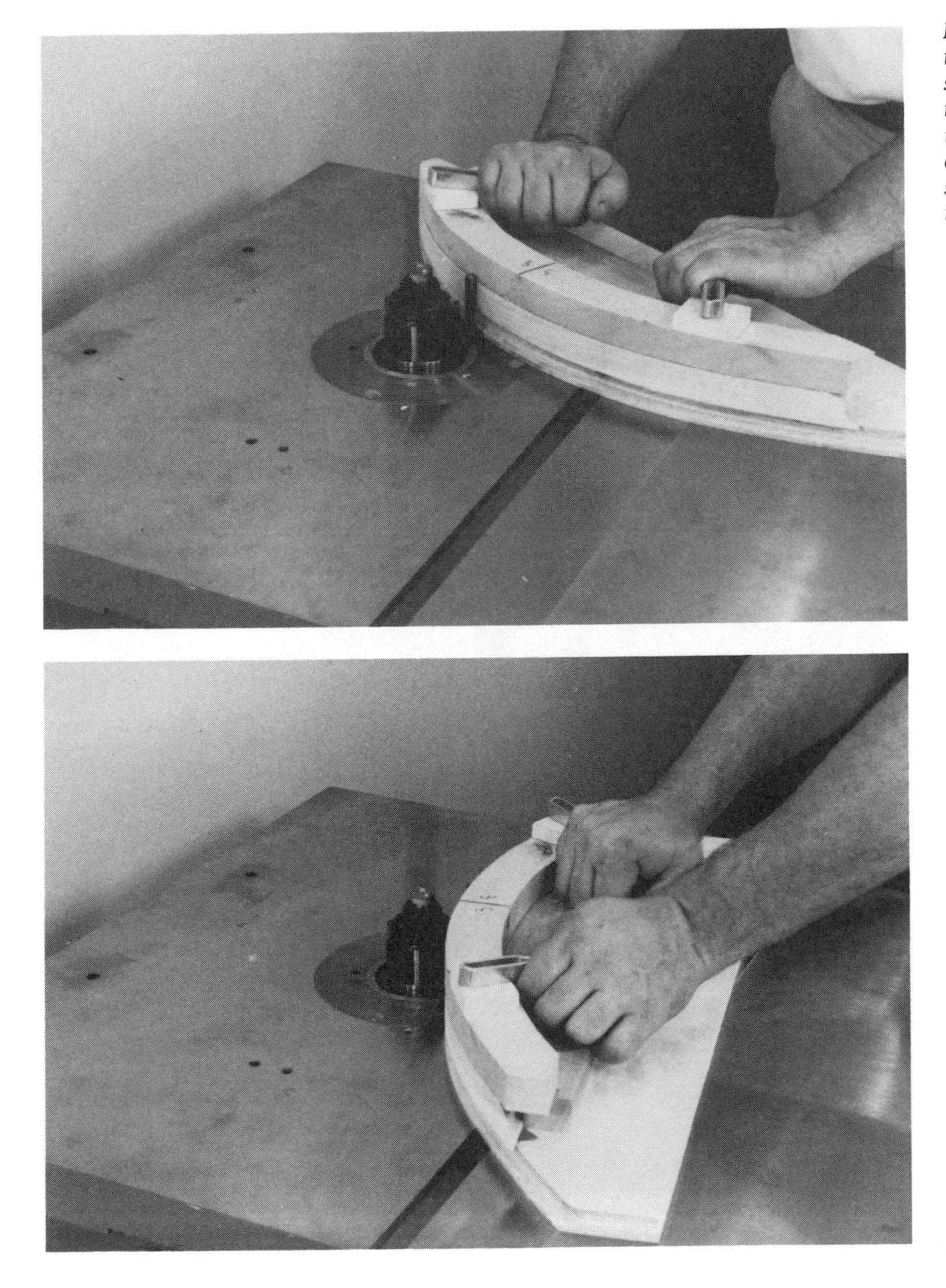

Illus. 8-42. Position the template against the starting pin. As you turn the template, the cutter will begin on the centerline. Note that the starting pin is on the thrust side of the cutter.

Illus. 8-43. Once the template engages with the rub collar, pull it away from the starting pin.

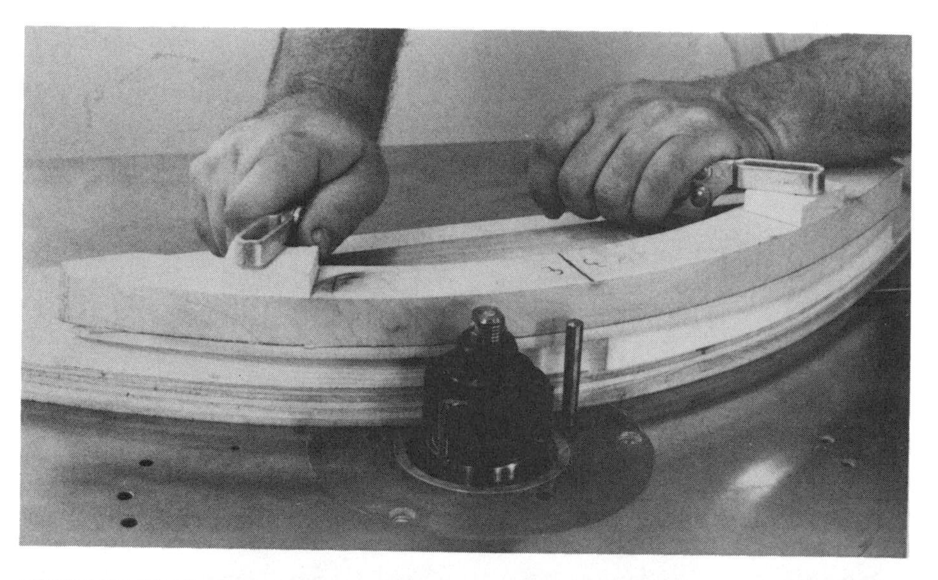

Illus. 8-44. Form the rabbet with a straight cutter. Comb the grain downward to control tear-out.

Illus. 8-45. Complete the cut with the template riding along the depth collar.

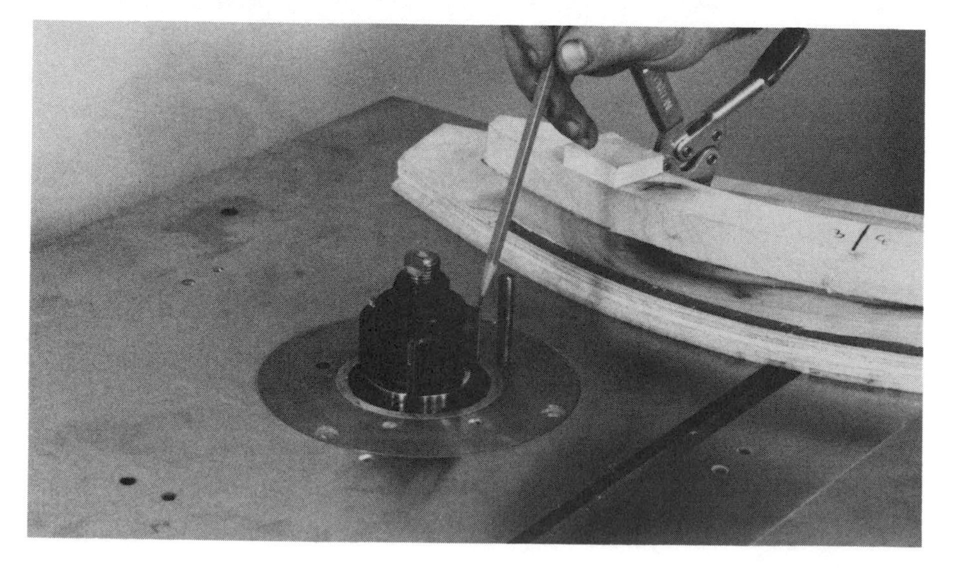

Illus. 8-46. Turn the straight cutter over to cut the rabbet on the other side of the moulding. Note: Reverse the rotation of the spindle before shaping the rabbet.

To shape the opposite side of the moulding, turn the cutter over (Illus. 8-46). This will require that you reverse the rotation of the spindle. You will also have to reposition the starting pin on the thrust side of the cutter (Illus. 8-47).

You can begin shaping in the rabbeted area, since no cutting will take place (Illus. 8-48). Cut until you have completed shaping the other side of the moulding strip (Illus. 8-49). This completes the glass mouldings (Illus. 8-50). Note that the rotation of the cutter and feed direction have changed. Cut the mitres and install the pieces.

To successfully shape the short-grain mouldings, you will have to reverse the spindle shaper. Absolute control of the work is essential. If the parts are not controlled, they may be fractured and ejected from the jig or template. For best results, take light cuts and maintain control of the work.

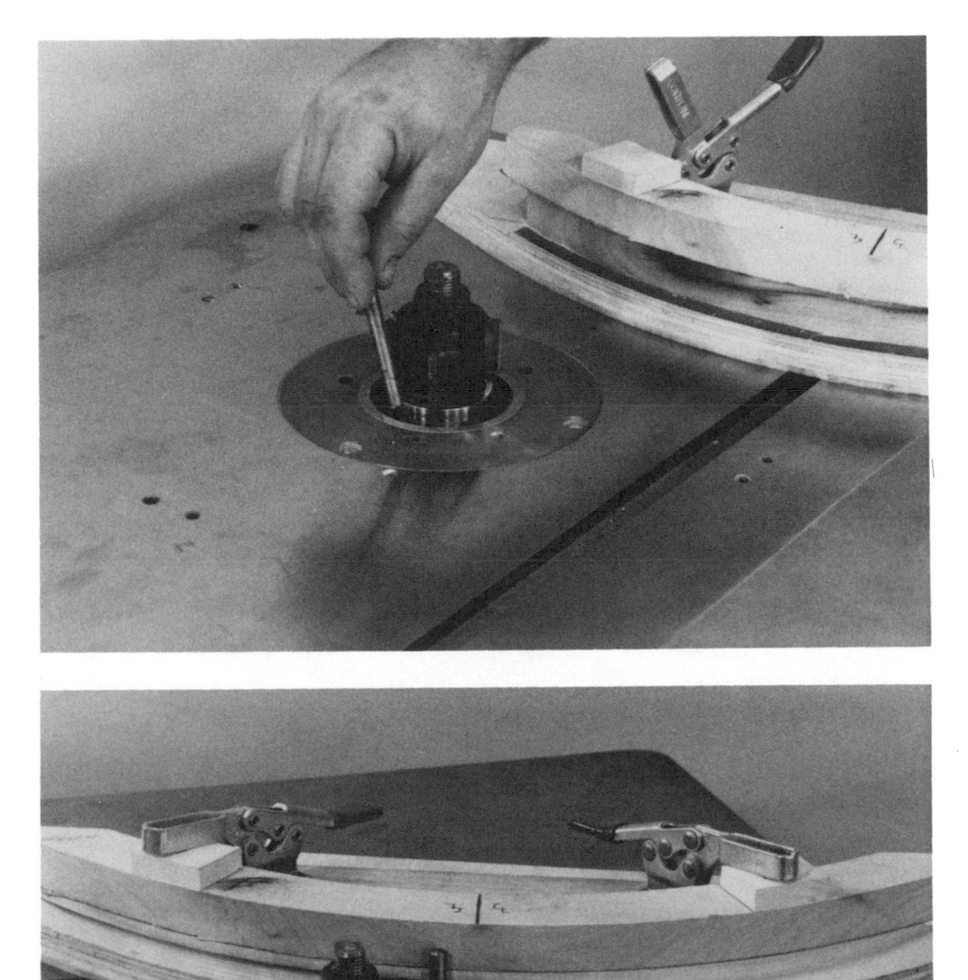

Illus. 8-47. Move the starting pin to the thrust side of the cutter.

Illus. 8-48. You can begin shaping in the rabbeted area, since no cutting will take place.

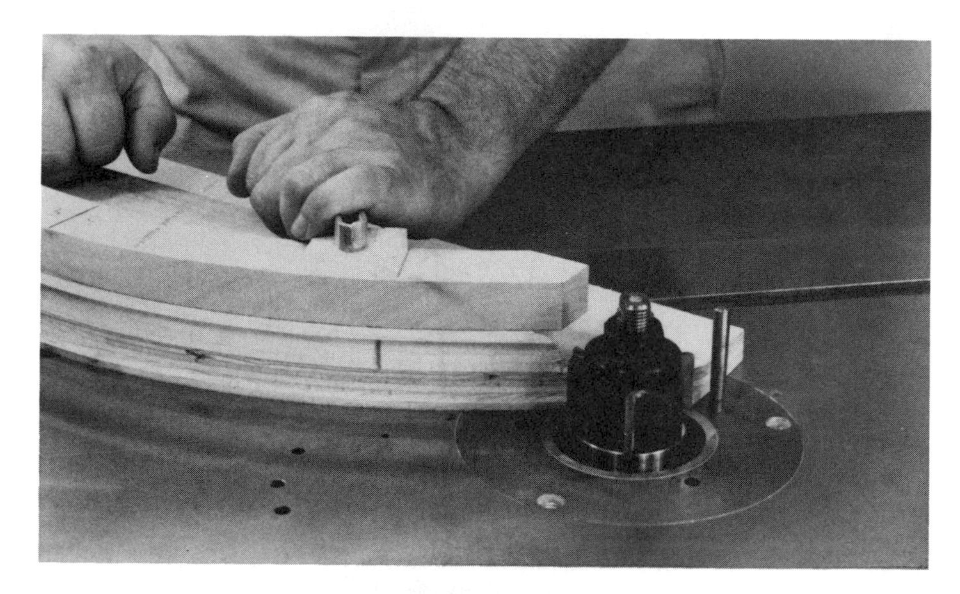

Illus. 8-49. Shape the other side of the rabbet. Comb the grain downward.

Special Power-Feed Setups

Some large mouldings can only be shaped safely with a power feeder (Illus. 8-51). The problem with power feeding this type of moulding is that the entire control edge is shaped away. This could cause the moulding to be twisted as the power feeder exerts pressure on the outfeed side of the moulding.

To make this moulding using a power feeder, change the profile of the outfeed fence (Illus. 8-52). Secure the radiused piece to the fence. It supports the work as the power feeder guides the work through the shaper.

Curved parts like the one on the left in Illus. 8-53 cannot be shaped horizontally on any shaper. That is because the profile is highest in the center. The shaper cutter cannot be ground for this operation.

A vertical profile could be used to shape this moulding, but a curved table would be required (Illus. 8-54). The difficulty of this operation is the large cutter and deep profile. This makes hand-feeding unsafe.

To make this curved moulding, adapt a power feeder to the job (Illus. 8-55). Use special metal

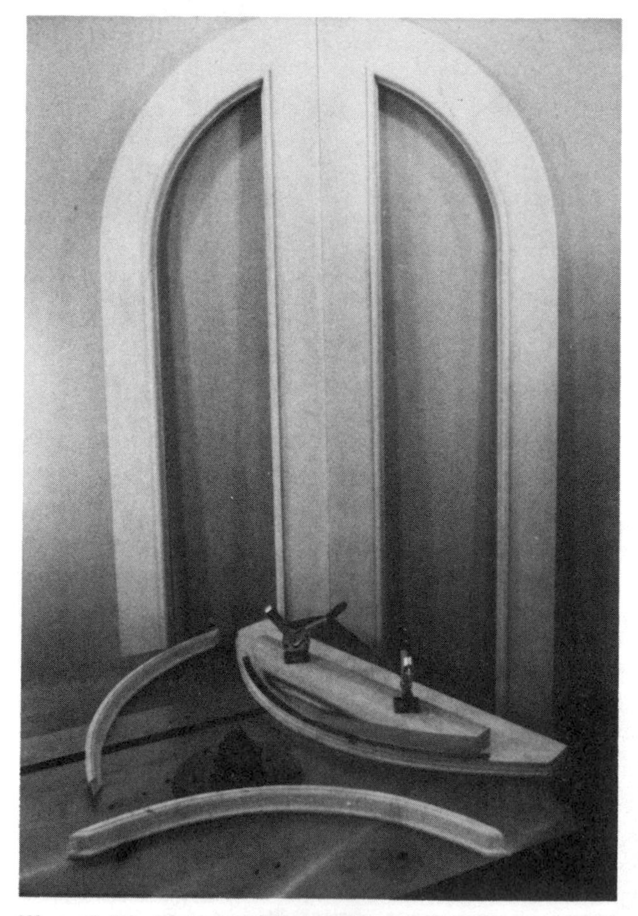

Illus. 8-50. The mouldings are now ready to be mitred and fitted into the doors.

Illus. 8-51. To safely shape large mouldings in one pass, you should use a power feeder.

Illus. 8-52. After shaping the moulding, change the outfeed fence profile to support it.

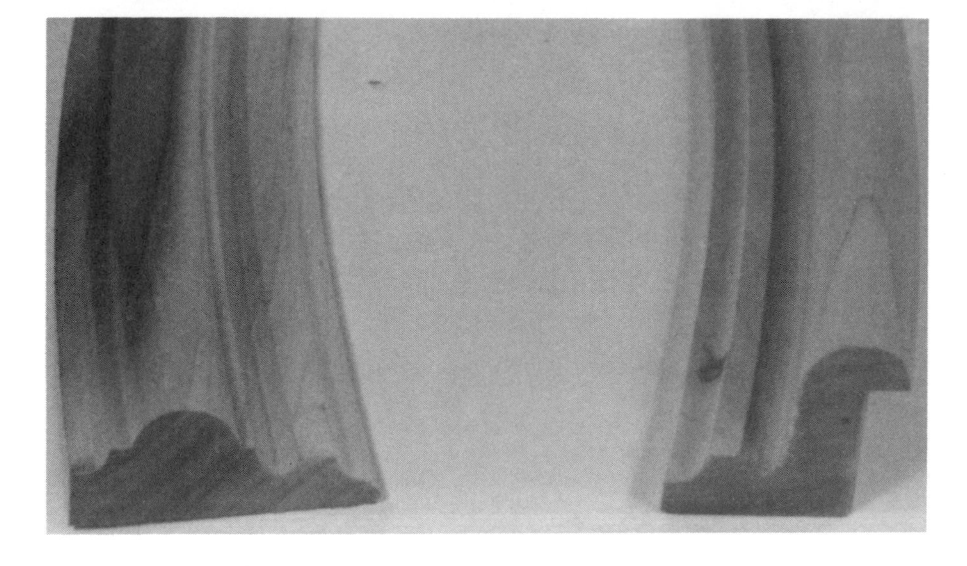

Illus. 8-53. The curved moulding on the left requires a vertical-profile cutter.

Illus. 8-54. A curved auxiliary table is secured to the fences. Note how the fences have been coped.

Illus. 8-55. A power feeder has been adapted for use with this setup.

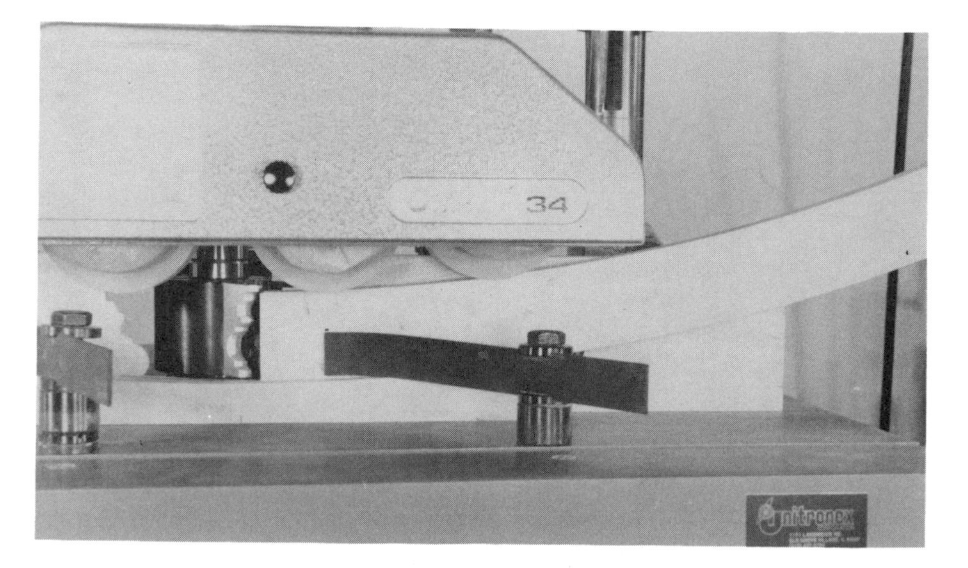

Illus. 8-56. The metal hold-downs keep the work against the fence. Make sure that they are clear of the cutter's path.

Illus. 8-57. The wheels of this power feeder deflect to feed the workpiece into the cutter.

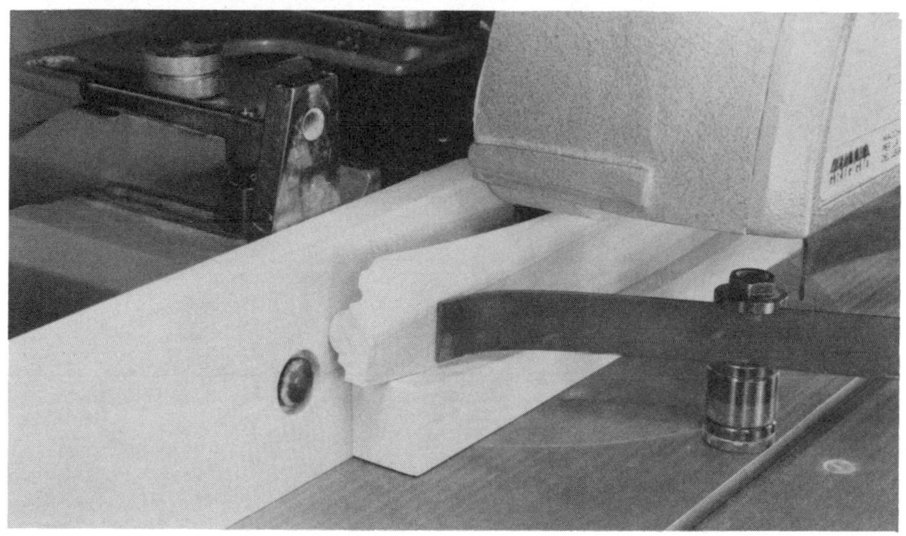

Illus. 8-58. This approach to shaping curved mouldings makes the operation safer and produces quality moulding.

hold-downs to keep the work against the fence (Illus. 8-56). The wheels of the power feeder deflect to feed the workpiece (Illus. 8-57). The power feeder should feed the workpiece at about 15-25 feet per minute. The quality of the moulding is excellent, and the operation is much safer than it would be if the work were hand-fed (Illus. 8-58).

Using Router Bits for Shaping

Using router bits mounted in the shaper allows you to make a wide variety of freehand and straight pattern cuts at a more reasonable cost. The shaper at this point will take the place of a handheld router. All you have to do is remove the spindle and replace it with a ½-inch collet. Any type of router bit can be used with a ½-inch-shank diameter.

Using a router bit with a ball-bearing pilot tip eliminates the need for a rub collar (Illus. 8-59). You will still have to use a starting pin or adjustable infeed fence.

When using a shaper with router bits, you can change the rpm of the machine for different size bits and diameters to produce a more uniform cut for the wood being used. You'll find that the workpieces that were a problem to clamp while you were using a router can now be held by hand and moved through the machine.

Illus. 8-59. Router bits can be used with a special collet adapter. A ball-bearing pilot tip acts like a rub collar. Always use a fence or starting pin with a router bit.

Illus. 8-60. Corian can be edge-shaped the same way wood is. Be sure to use carbide cutters.

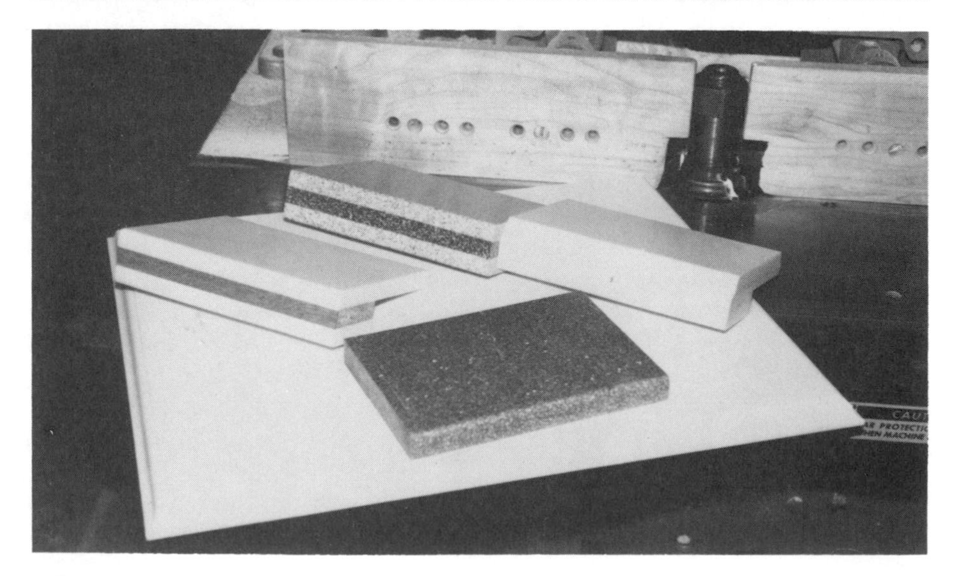

Illus. 8-61. Specialty edges and joinery can also be cut in Corian or wood and Corian.

The shaper collet can rotate clockwise or counterclockwise to prevent any problems with grain tearout. Make sure that the rotation of the shaper is correct for the bit.

Today, most shaper cutter profiles can also be made with smaller sized router bits. Do not use bits with a ¼-inch shank. The force against the bit may be sufficient to break the bit.

Using a router bit collet for your shaper will prove economical because router bits are cheaper than shaper cutters. It will also make your shaper more versatile, because you will be able to use shaper cutters *and* router bits in your setup.

Router bits have a smaller diameter than most shaper cutters. This means that their tips or peripheral speeds will be lower at any given revolution per minute. For effective shaping with router bits, you may have to increase the speed of the spindle.

It is good practice to measure the shank of the router bit before using it. If it is undersize, it may not fit the collet securely. Select high-quality bits for use in the shaper collet.

Shaping Corian™

Corian™ is a synthetic material that looks like marble. It is quite heavy, and is favored for countertop work. Corian can be worked with carbide tools and cutters. Special radii and edges can be cut with the shaper (Illus. 8-60). Specialty edges and glue joints are cut the same way wood is cut (Illus. 8-61).

The weight of Corian makes it advisable to use a power feeder. Cutting speeds for Corian are similar to those of wood. Look for signs of high cutting or feed speed, and make needed adjustments.

About the Authors

Roger W. Cliffe is the author of four books: *Woodworker's Handbook, Table Saw Techniques, Radial Arm Saw Techniques,* and *Portable Circular Sawing Machine Techniques.* Cliffe, who has taught woodworking for the past twenty years and maintains a small specialty shop in his hometown, has also produced a series of videotapes on cabinetmaking. He is a professor in the Department of Engineering Technology at Northern Illinois University in DeKalb, Illinois.

Michael J. Holtz comes from a woodworking family—his brothers and father own a business devoted to commercial displays. Holtz's love of fine woodworking caused him to take another path. He owns a woodworking business known as *Der Holtzmacher* (German for the woodworker), specializing in church interiors. This accounts for Mike's vast knowledge of the shaper and shaper setups. He designs church interiors (in consultation with architects), and then builds and installs them. Holtz, Cliffe's student in the early 1970s, opened *Der Holtzmacher* after teaching woodworking and selling wood crafts for five years at art fairs.

Authors Roger Cliffe (left) and Michael Holtz

Index

T

U

V

W